NATIONAL LIBRARY OF AUSTRALIA

A catalogue record for this book is available from the National Library of Australia

ISBN (print): 978-1-925823-47-9
ISBN (e-book): 978-1-925823-46-2

Contents

**3.   Anti-biofouling by Hydrophilic Polymer Brushes   61
     and Force Measurement of Cypris**
*Motoyasu Kobayashi, Yuka Yamaguchi and Shouhei
Shiomoto*

**4.   Hybrid Organosilicone Materials as Efficient Anti-   81
     corrosive Coatings in Marine Environment**
*Rami Suleiman, Amjad Khalil, Mazen Khaled and
Bassam El Ali*

**5.   Polymers and Polymer Composite Coatings for   115
     Marine Applications: A Review**
Swati Singh and Vikas Mittal

Preface

Marine biofouling on surfaces is a global problem with detrimental effects. Some of the common issues associated with biofouling include higher fuel consumption for marine vessels, corrosion, emission of greenhouse gases, etc. Biofouling is a complex biochemical phenomenon due to the involvement of a vast diversity of fouling organisms, thus, making the prohibition very challenging. Coatings represent one of the important strategies to achieve protection against biofouling and its effects due to the incorporation of novel materials as matrices. In this respect, the book aims to present many functional marine coatings systems developed in the recent years to achieve effective protection against biofouling. Recently, membrane technology has also gained considerable interest because of simple and environmentally friendly approach for seawater treatment. Thus, the book also focuses on the recent developments gained in this area.

Chapter 1 reports that the hybrid self-stratified siloxane-polyurethane coatings show great potential for use in practical marine applications. Siloxane chains migrate to the coating-air interface upon crosslinking, thus, providing a hydrophobic character to the coatings. In Chapter 2, the effect of residual solvent and temperature on the corrosion resistance of high performance polymers was studied by fabricating polyetherimide coatings on carbon steel substrate. Chapter 3 describes the synthesis of polymer brushes and their biofouling properties preventing the settlement of cypris larvae in seawater. In addition, the recent studies on the adhesion force measurement of the live cypris and footprint protein are also described. Chapter 4 reports that the hybrid sol-gel materials result in efficient protective anticorrosion/antifouling coatings for metal substrates in the marine environment. The field trial results of the case study reported in this chapter prove that the presence of both corrosion inhibitors and protective antifouling bacteria within the network of a sol-gel coating can significantly reduce the antifouling properties comparing to a coating containing protective bacteria alone. Chapter 5 reviews the use of polymer composite coatings for various marine applications. In Chapter 6, particular attention has been devoted to the description of the latest approaches for the chemical modification of polydimethylsiloxane to render the surface more hydrophilic and the entire coating amphiphilic, with an aim to enhance its antifouling activity and fouling-release capacity, thus, improving its performance when applied to vessels that cruise at relatively low speeds or spend idle periods in port. In Chapter 7, the development of stable chitosan coatings on mild carbon steel has been reported by layer-by-layer (lbl) addition of chitosan and hydrophobic polymer like poly(vinyl butyral), along with the analysis of the anti-corrosion performance in 0.3 M salt solution. Chapter 8 comprehensively discusses polymeric marine membranes for water (seawater and contaminated water) remediation in general and strategies to impart antifouling and antibacterial properties in particular.

The book would not have been successfully accomplished without the support of chapter contributors. The book is dedicated to my family for unswerving support and constant motivation.

Vikas MITTAL

1

Self-stratified Siloxane-Polyurethane Fouling-release Marine Coating Strategies: A Review

Madhura Pade and Dean C. Webster*

Department of Coatings and Polymeric Materials, North Dakota State University, Fargo, ND 58108 USA

Corresponding author: dean.webster@ndsu.edu

1.1 Introduction

Marine biofouling is the dynamic process of attachment and growth of aquatic organisms, such as algae and barnacles, on surfaces submerged in natural water bodies [1]. Although highly dynamic, the process of biofouling can be described in four main stages: formation of a conditioning layer of proteins and polysaccharides, attachment of bacteria, settlement of diatoms and microalgae, and finally attachment of macrofoulants like barnacles and mussels. Over the years, biofouling has proven to be a nuisance for man-made structures resulting in environmental and economic disadvantages. Common disadvantages for marine vessels include higher frictional drag and fuel consumption, difficulty in maneuvering ships and vessels, corrosion of the underlying substrates, emission of greenhouse gases, transfer of aquatic species to non-native environments, etc. Not just structurally and environmentally, the economic impact of biofouling cannot be ignored either. For example, based on the type of marine coatings, the current hull cleaning practices and the level of fouling, the estimated overall cost of cleaning ship hulls of the Arleigh Burke DDG-51 destroyers (30% of the ships in the US Navy fleet) is ~$56 million US dollars annually [2-3]. For the entire naval fleet, the costs can increase up to ~$220 million per year [2]. The complexity of the biofouling phenomenon makes it difficult to identify a single practical solution to combat the attachment of aquatic species.

Combating biofouling has been extremely challenging due to the identification of more than 4,000 different marine organisms, each

Marine Coatings and Membranes, edited by Vikas Mittal
© 2019 Central West Publishing, Australia

exhibiting different attachment behaviors. Historically, wooden ships were covered with lead, tar, wax, asphalt and copper [1,4]. After replacement of wood with iron, copper or lead sheathings were used as antifouling layers [1]. The need for sophisticated technologies to combat biofouling led to the introduction of antifouling (AF) coatings. AF coatings contain tin or copper based biocides, dispersed in various binders [1,5]. However, the long term toxic effects of the biocides leached from ship hulls [6] has necessitated further research into exploring safer alternatives to combat biofouling. First described in the 1970s, fouling-release (FR) coatings are made using low surface energy (SE) materials and do not contain toxic biocides [4,7]. The low surface energy of the coating is expected to cause the fouling organism to be easily removed from the surface [1,4]. If organisms attach to the coating, hydrodynamic forces from movement of the vessel can easily overcome the weak interactions between the organisms and the surface [1,5]. Initial results with FR coatings indicated subpar performance of the coatings as compared to the tributyl tin self-polishing AF coatings [1,5,7]. Since the 2000s, research to develop novel FR coatings, which reduce interactions between the surface and the organism, has gained momentum [5]. Due to considerations of surface energy, surface porosity and fracture mechanics, siloxane elastomers have become the system of choice for fouling release coatings [8-11]. Silicone oils are often added to provide an interfacial slip layer at the coating surface. A number of commercial companies provide FR coatings which are based on silicone elastomers. Although the coatings exhibit good FR behavior, the elastomers are susceptible to damage due to their low modulus and low strength. Furthermore, the non-wetting nature of silicones makes adhesion of silicone elastomer coatings onto substrates such as marine epoxy primers difficult. Even if a tie coat is applied prior to application of the silicone elastomer, the adhesion of a silicone elastomer to the tie coat may not be ideal for practical applications. To overcome the disadvantages of silicone elastomer based coatings, the concept of the self-stratified siloxane-polyurethane (SiPU) coating system was developed. Illustrated in Figure 1.1, the coating system consists of a two-component polyurethane made using a polyol and a polyisocyanate to which a functional polydimethylsiloxane (PDMS) is added, which can react with isocyanates. During film formation, the siloxane component spontaneously segregates to the surface due to its low surface energy, while the bulk of the coating is polyurethane, yielding good toughness and adhesion to the substrate.

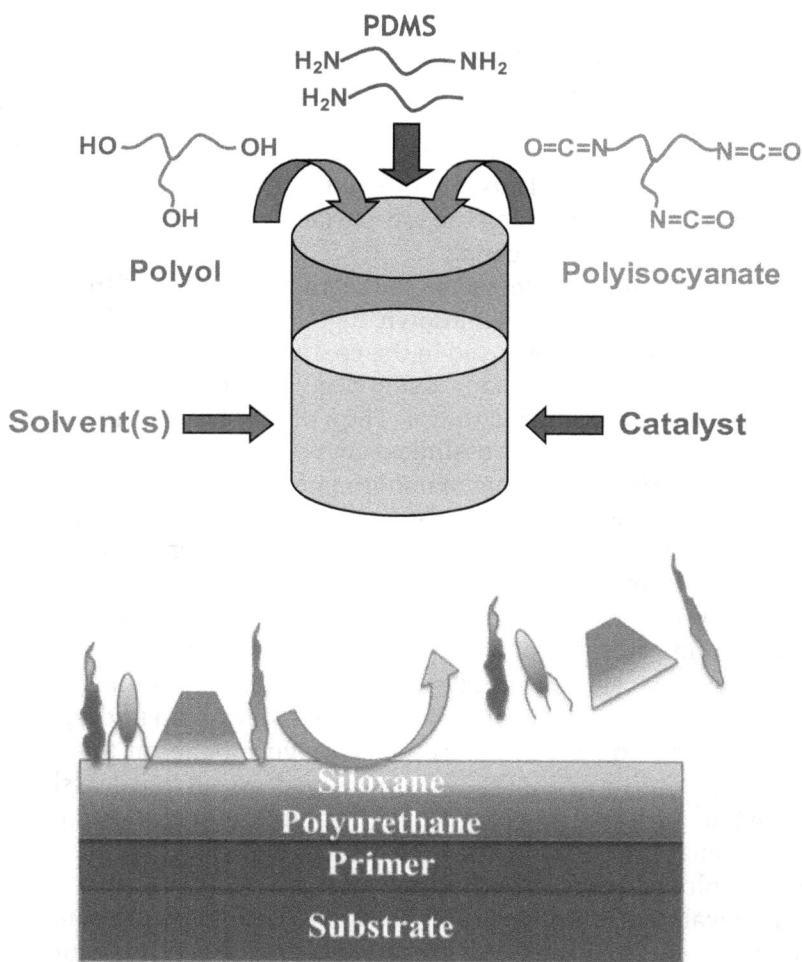

Figure 1.1 Components of the siloxane-polyurethane fouling release coating system (top). Schematic of the self-stratified coating after application (bottom).

1.2 The SiPU Hypothesis and Proof of Concept

This approach was inspired by the numerous studies in the literature that indicated that if a film of a block copolymer of PDMS and any other polymer is formed, the PDMS phase will predominate on the surface, yielding a low surface energy, while the bulk properties will be dominated by the other polymer [12-18]. The primary driving force for this surface enrichment is surface energy minimization, and, since PDMS has a much lower surface energy than other organic polymers, it predominates on the surface. In fact, very little PDMS in the overall system is needed for PDMS to essentially provide complete surface coverage [19]. Thus, we desired to utilize this phenomenon to generate a coating having the low surface energy needed for a marine fouling release application.

Polyurethane was selected as the main component of the coating system since it is a very well-known, highly durable and high performance technology widely used in the coatings industry. Two-component polyurethanes are made using a polyol and a polyisocyanate which are mixed just prior to use. They then react on the substrate and cure to form a highly crosslinked coating film. Polyurethanes also develop an extensive hydrogen bonding network, which gives rise to the toughness and abrasion resistance found in these systems. Polyurethane coatings are used for aircraft, vehicles, bridges, chemical storage tanks, process equipment and many other applications.

Siloxane-polyurethane block copolymers have been explored for use as fouling-release coatings [20-21]. However, while these copolymers have low surface energy surfaces suitable for fouling-release, under prolonged immersion in water, the material can undergo rearrangement to express the more polar polyurethane on the copolymer-water interface [22-23]. Thus, we hypothesized that crosslinking the system during curing in air could "lock in" the stratified morphology and enable the PDMS phase to remain expressed on the surface under prolonged exposure to water.

A typical two-component polyurethane coating system consists of a polyol, a polyisocyanate, curing catalyst, pot life extender and a solvent package. Using a multifunctional polyol and/or multifunctional polyisocyanate will result in a crosslinked thermoset. For the siloxane component, PDMS having organofunctional end groups, such as alkyl amine or alkyl alcohol, is incorporated into the formulation. After mixing, when applied to a substrate in a single step, the components are expected to spontaneously phase separate into distinct layers. Self-stratification is expected to occur due to differences in surface energy and solubility parameters of the components, along with the viscosity of the system. Therefore, a combination of non-polar PDMS and polar PU components was explored in anticipation that the resultant coating would show self-stratification. The PDMS chains would form the low surface energy top layer, while the polyurethane component will provide adhesion to the substrate as well as tough bulk properties.

A challenge that is faced in this process is the large number of variables present in the coatings system, most of which could be expected to have some impact on the degree of stratification and overall properties of the coatings. For example, polyols of different type (polyester, polyether, acrylic), composition, molecular weight and functionality can be used. PDMS having different end groups and molecular weights could also be used in the system. Further, variables such as different polyisocyanates, catalysts, catalyst amounts, solvents having different evaporation rates and solubility parameters, etc., could all potentially affect the properties of the coatings. To be able to address the complexity of the variable space in this system, we employed combinatorial as well as high throughput methods and constructed a highly automated laboratory that could enable the rapid preparation and screening of hundreds of polymers and coating compositions [24-36]. These methods proved to be invaluable in the design and optimization of the compositions to achieve coatings having good fouling-release properties.

The initial experimental study explored the effect of formulation variables (isocyanates, solvents, etc.) on coating properties and changes in properties of the cured coatings (contact angle (CA), pseudobarnacle adhesion (PBA)) upon exposure to an aqueous environment. Preliminary experiments were conducted to incorporate PDMS oligomers with alcohol or amine functional groups, polyols, catalysts, solvent mixtures and pot life extenders in coating formulations. Often, poor compatibility between PDMS and other components was evident from formation of two distinct phases upon removal of agitation or stirring. When the phase separated formulations were cured, the resultant coatings showed significant phase separation of the PDMS and polyurethane components. Therefore, determination of best solvents or solvent blends was identified as one of the key considerations for the SiPU formulations.

Experiments were carried out comprising four different coating libraries, for a total of 82 formulations [37]. The formulations were made using two different isocyanate crosslinkers, polycaprolactone triol (PCL), hydroxy alkyl terminated (10k g/mol) and aminoalkyl functional PDMS (APT-PDMS), catalyst dibutyltin diactetate (DBT-DAc) and 2,4-pentanedione pot life extender (Figure 1.2). Two APT-PDMS, with MW 12k g/mol and 24k g/mol, were evaluated. All formulations were applied using drawdown and cured at 80 °C for 1 hour to give coatings with 25-50 μm dry film thickness. Water contact angles (WCA) and surface energy were determined on both as-made coatings and after immersion in deionized water to assess the surface properties and the stability after water immersion. Pseudobarnacle adhesion (PBA) strength of the coatings was determined using an automated adhesion tester to simulate adhesion of barnacles using aluminum studs that were attached onto the coatings using an epoxy adhesive [31]. Dibutyl tin diacetate was selected as the catalyst over the more popular dibutyl tin dilaurate since it is more water soluble and easier to leach from the coating prior to carrying out laboratory biological assays. Since the tin catalysts are biocides, being able to rapidly leach them from the films is critical to obtain accurate fouling-release results.

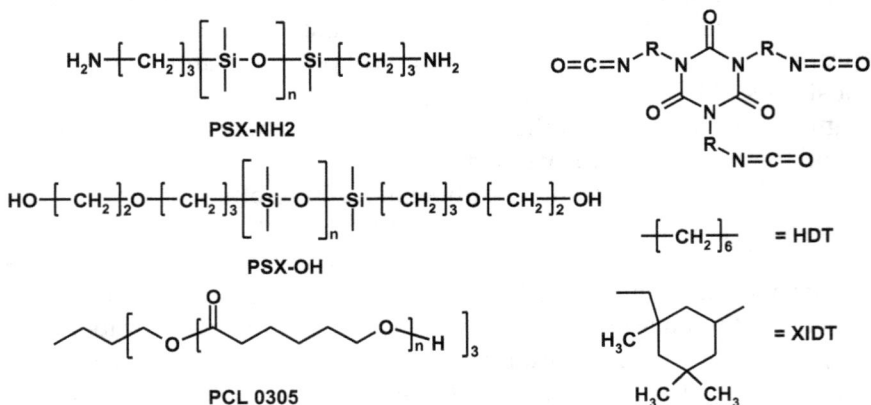

Figure 1.2 Components used for the first study with the self-stratifying SiPU coatings. Reproduced from Reference 37 with permission from American Chemical Society.

The first formulation library comprised of 10-20% APT-PDMS, PCL and isophorone diisocyanate (IPDI) isocyanurate trimer ("XIDT"). Since amines are more reactive with isocyanates than alcohols, using an amino alkyl functional group in the PDMS was expected to ensure complete chemical reaction of PDMS into the coating network upon mixing of the components. This library also served as a starting point to explore common solvents, such as methyl amyl ketone (MAK), butyl acetate (BA), isopropyl alcohol (IPA), ethyl 3-ethoxy propionate (EEP), toluene and xylenes. It is common in coating systems to have a mixture of different solvents in order to control the film formation process. WCA experiments prior to water immersion showed that the coatings formed hydrophobic surfaces. Most of the formulations remained hydrophobic after 30 days of water immersion, although a drop in WCA was observed for some formulations. Furthermore, coatings made using EEP and an EEP:toluene mixture showed the lowest PBA force among the different formulations, indicating that the solvent can play a role in establishing the surface properties of the coatings.

For the second library, IPDI trimer was replaced with a hexamethylene diisocyanate (HDI) isocyanurate trimer ("HDT"), while retaining the other components. Similar to the first library, most of the formulations with HDI trimer formed hydrophobic coatings initially and after water immersion. Formulations made using EEP and an EEP:toluene mixture again resulted in coatings having the lowest PBA force.

A third library was then prepared to determine the effect of PDMS content on the properties of the cured coatings. In the third library, formulations were made using two crosslinkers - IPDI and HDI trimers, the two best solvents - EEP and toluene, and APT-PDMS at different levels. For the third library, WCA values were measured weekly for 4 weeks. The results showed that the coatings maintained their hydrophobicity after exposure to water (Figure 1.3).

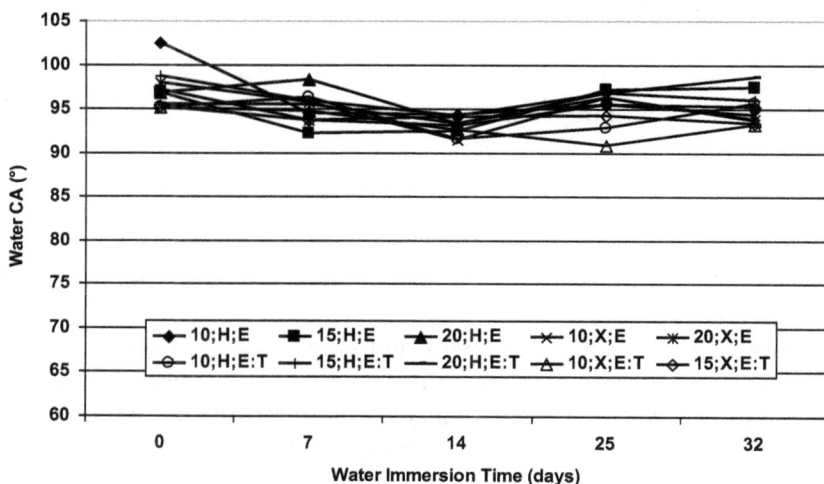

Figure 1.3 Water contact angle data for siloxane-polyurethane coatings as a function of water immersion time. Data series are given as percent PDMS:isocyanate (H=HDT; X=XIDT):solvent(E=EEP,E:T=EEP and toluene blend). Reproduced from Reference 37 with permission from American Chemical Society.

In the fourth library, the hydroxy alkyl terminated PDMS was used in place of the APT-PDMS. Similar to the other libraries, most formulations formed hydrophobic coatings prior to water immersion. From these results, it was concluded that using the amino alkyl terminated PDMS, coupled with the proper mixture of solvents, resulted in coatings having stable hydrophobic surfaces.

This comprehensive first work was the evidence that stable self-stratified crosslinked coatings can be made using functional siloxanes, isocyanates and polyols. The study proved useful in identifying important variables that play an important role in properties of the resultant coatings. Moreover, analysis techniques such as CA/SE measurements and PBA adhesion tests proved vital in gaining preliminary understanding of the coatings, and these tests were included for most of the future studies.

1.3 Coatings with Surface Microdomains

In another study, a series of coatings were made using hydroxy functional PDMS (1k g/mol), PCL and IPDI polyisocyanate. PDMS content was varied as 0 (polyurethane control), 10, 20 and 30% solids [38]. Characterization of the cured coatings using atomic force microscopy (AFM) showed that the formulations with 20 and 30% PDMS as well as the PU control formed smooth and domain-free coatings. Formulations containing 10% PDMS formed surfaces with distinct domains, 1.4 μm diameter and ~50 nm high. The results from nanoindentation experiments revealed that the microdomains possessed lower modulus than the matrix, indicating PDMS as the primary component in the domains, while the matrix was primarily PU. Analysis using scanning electron microscopy (SEM) and energy-dispersive X-ray mapping of silicon (Si mapping) showed evidence of formation of PDMS-rich surface domains. However, SEM imaging and Si mapping of the coating with 30% PDMS showed that silicon was uniformly distributed over the surface. To determine if the domains were formed from unreacted PDMS, the coating was treated with toluene, which would have dissolved away any unreacted PDMS. AFM analysis of the treated coating showed that the domains swelled in toluene but could not be removed from the coating. The coatings were placed in water for 2 weeks to determine coating stability in aqueous environment. Changes in size and distribution were observed after water immersion, with the diameter increasing from 1.4 μm before water aging to 3.82 μm after water aging.

Phase separation of the PDMS chains is believed to be dependent on a number of factors, such as SE differences, incompatibility due to differences in solubility parameters and flexibility of siloxane chains. During the later stages of the reaction, a dynamic ordering process causes changes in the morphology of the coatings possibly due to changes in reaction kinetics. Therefore, apart from solvent composition, the kinetics of the reaction during mixing and film formation are expected to directly affect the surface microstructure of the cured coatings. Therefore, in another study, the role of solvent composition as well as mixing time and catalyst loading in controlling the formation of a microtopographical coating was studied using a statistical experimental design approach to better understand the nature of the spontaneously biphasic hybrid SiPU coating system [39]. The solvent composition provides initial compatibility to the precursor oligomers. In addition, both the polarity and evaporation rate of the solvent can dictate the rate of curing and initiation of phase separation as the solvents evaporate and the coating cures on the substrate. To study the effect of solvents on the nature of the domains formed, MAK, toluene, EEP, BA and IPA were used as solvents to make 35 different formulations. All the formulations contained 10% PDMS-1k and IPDI polyisocyanate. AFM analysis of cured coatings showed that the solvent composition determined not only the formation of the surface domains, but also the size of the domains (Figure 1.4). A model was developed and correlated to the vapor pressure and solubility parameters of the solvents to quantify their effects. The results showed that to achieve domain formation, the vapor pressure of the solvent compositions should be in the range of 5.27-10.54 mm Hg and the solubility parameter from 8.56-8.62 $(cal/cm^3)^{0.5}$.

To follow up on an observation that mixing times are an important variable for phase separation [39], an extensive study was conducted to understand the effect of mixing time of the formulations on the morphology and surface properties of the cured SiPU coatings [40]. The study involved a single formulation, comprising 10% PDMS-1k, PCL polyol, IPDI-based polyisocyanate and 0.075% DBTDAc catalyst. The components were mixed and drawdowns were made at 0.5-1-hour intervals over 7 hours. Solvent blend of EEP:MAK:BA = 43:12:45 was used as the solvent composition in all the formulations. Between 3-6.5 hours of mixing time, domains were formed with sizes varying from 2.09-0.84 µm diameter and 85.4-24.9 nm height, with both the height and diameter decreasing with mixing time. Interestingly, surface analysis using contact angle measurements demonstrated little difference among the coatings, regardless of domain formation. Barnacle adhesion was also determined using a laboratory barnacle re-attachment assay [40,41]. Coatings having surface domains showed low barnacle adhesion values (Figure 1.5).

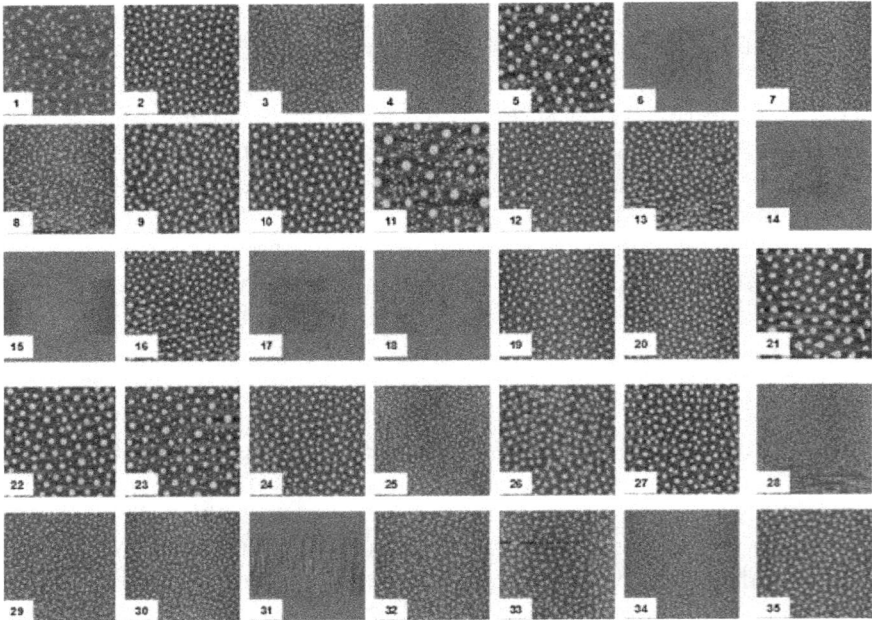

Figure 1.4 AFM high images showing presence or absence of surface domains as a function of solvent composition. Reproduced from Reference 39 with permission from Elsevier.

Another study was conducted to determine the effect of content and MW of PDMS, duration of mixing and method of mixing on the formation of a micro-topographical surface [42]. The extent of the isocyanate-hydroxyl PDMS reaction was determined using FTIR. Monte Carlo simulations were used to identify the degree of conversion at which phase separation occurs, leading to microdomain formation. The study revealed a "window" of reaction time that led to the formation of surface microdomains. Results from this study and the previous experiments showed that the microdomains can be precisely tuned by altering coating formulations. While some initial results showed that the coatings having surface microdomains could reduce the adhesion of barnacles, it was found that other smaller organisms such as bacteria and algae could easily colonize the surface

with poor release properties [43]. Thus, this approach, while resulting in interesting surface morphology, was not pursued further for
marine fouling-release coatings.

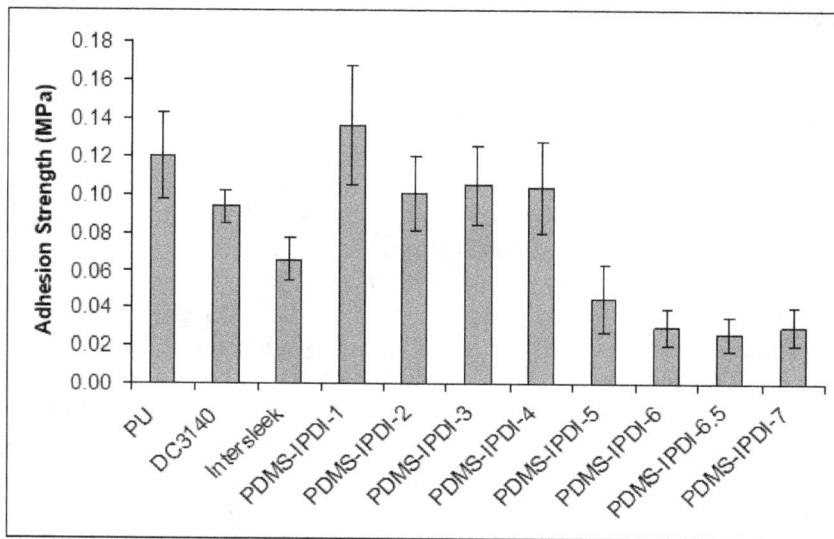

Figure 1.5 Adhesion strength of barnacles on siloxane-polyurethane
coatings and control coatings polyurethane (PU), DC3140 (silicone
elastomer), and Intersleek (marine fouling-release coating). Reproduced
from Reference 40 with permission from Federation of Societies for
Coatings Technology and Oil and Colour Chemists' Association.

1.4 Effect of Composition Variables on Properties of SiPU
Coatings

The initial set of experiments were essential in establishing some of
the basic parameters of the siloxane-polyurethane coating systems,
specifically, the solvent composition, choice of catalyst and amount as
well as the use of a pot-life extender [37]. Following this, a series of
studies were carried out to explore the composition variables and
their effect on surface and fouling-release properties in more detail.
The majority of these studies were carried out using combinatorial
and high throughput experimentation (C/HTE) in order to enable the
preparation of a large number of compositions and determine their
properties rapidly [29,33]. A series of laboratory biological assays

was also developed in order to be able to study the fouling-release properties of the libraries of coatings using relevant marine organisms [44].

1.4.1 Effect of PDMS Composition and Molecular Weight

A study was designed to explore the effect of PDMS molecular weight as well as the nature of the chain end on the fouling-release properties of the coatings. Since a polycaprolactone (PCL) polyol was used in the polyurethane system, it was hypothesized that using PCL-PDMS-PCL triblock copolymers might affect the degree of compatibility and amount of phase separation achieved in the SiPU coating system, thus, impacting the surface properties. The synthesis involved the preparation of aminopropyl terminated PDMS (APT-PDMS) of varying molecular weights followed by ring-opening polymerization with ε-caprolactone to form the triblock copolymers. An initial study was carried out to establish the synthesis methods for APT-PDMS over a range of molecular weights and to synthesize the triblock copolymers in a well-controlled fashion using the high throughput synthesis reactor, so that a library of polymers could be created [45].

Using C/HTE methods, formulations were generated to understand the effect of siloxane component (APT-PDMS oligomers and PCL-PDMS-PCL copolymers) on the surface properties of the cross-linked SiPU coatings [46]. The molecular weight of the APT-PDMS was varied from 2.5 k to 35 k g/mol (six levels) and subsequently each of the APT-PDMS was endcapped with 2, 3 and 4 caprolactone units. This resulted in 48 different siloxanes for incorporation into the SiPU coatings. Formulations were made using different siloxanes at four different levels (10, 20, 30 and 40 wt%), PCL, IPDI trimer and DBTDAc catalyst. The coatings were then analyzed using WCA/SE measurements and PBA test before and after 30 days of water immersion to compare stability and changes in coating properties. Overall, a total of 192 coatings were formulated using 48 PDMS oligomers. The WCA values after water immersion are shown in Figure 1.6 for the entire set of coatings. Some interesting trends can be observed. First, the coatings made from the PCL-PDMS-PCL triblock copolymers essentially had WCA values greater than 90°, while those made using the APT-PDMS span a wide range of WCA values with many exhibiting values below 90°. Furthermore, it can also be observed that increasing the MW of the PDMS as well as the amount of PDMS polymer in the coatings resulted in a higher WCA value.

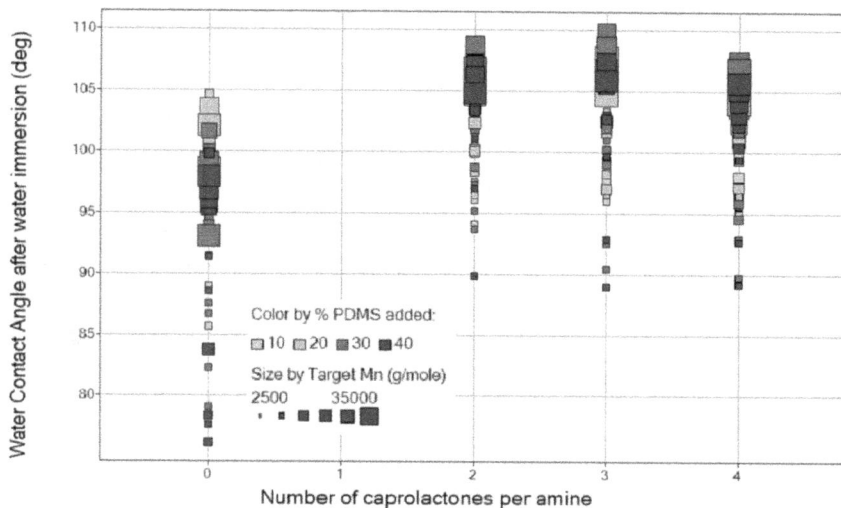

Figure 1.6 Water contact angle averages of 192 coatings after 30 days of water immersion vs. number of ε-CLs per amine. Size of the data points increases as the molecular weight of PDMS block increases, color of the data points gets darker with increasing PDMS content in coating formulations. Reproduced from Reference 46 with permission from American Chemical Society.

The PBA results for this set of coatings following water immersion are shown in Figure 1.7. The first interesting observation is that the values are low for the coatings made using APT-PDMS, in contrast to those made with the PCL-PDMS-PCL triblock copolymers. Low values are desired since they are an indication that the coatings may have good fouling-release performance. With the use of the triblock copolymers, PBA increases with the lowest adhesion values for the lowest amount of triblock copolymer in the coating system.

A selected set of coatings from this initial screening study were chosen for use in laboratory assays with marine organisms. The SiPU coatings consisted of the set using APT-PDMS over a range of molecular weights (PDMS-PU) and a set of PCL-PDMS-PCL triblock copolymers having three caprolactone units on each chain end (PDMS-PCL-PU). The copolymers were added to the SiPU formulation at 20%. In one study, the marine bacterium *Cytophaga lytica* (*C. lytica*) was incubated on the surface of the coatings and then stained with crystal violet for imaging [47]. In Figure 1.8, it can be noticed that the bacteria on the surface of the PDMS-PU coatings retracted from the coating

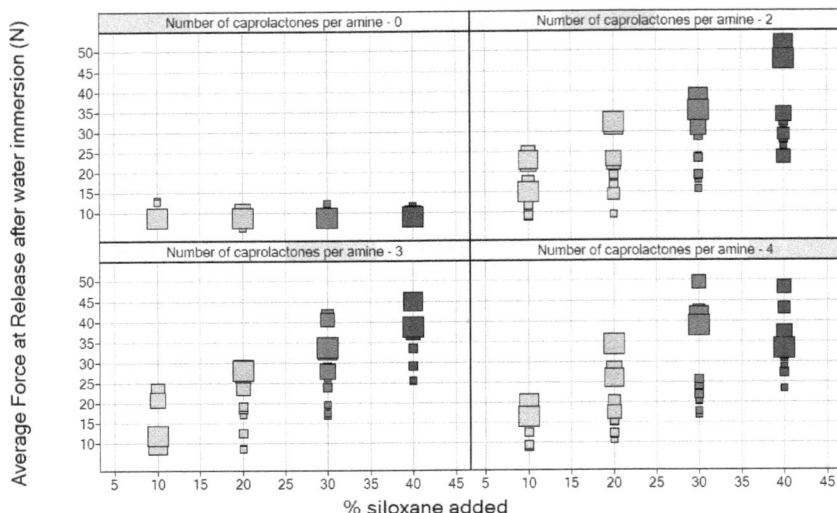

Figure 1.7 Pseudo-barnacle adhesion for SiPU coatings after 30 days of water immersion vs. amount of siloxane in coating formulations trellised by number of ε-CLs per amine. The size of the data points increases as target molecular weight of PDMS increases. The color of the data points gets darker as siloxane polymer level increases. Reproduced from Reference 46 with permission from American Chemical Society.

surface and formed clumps while the bacteria on the surface of the PDMS-PCL-PU coating was spread uniformly. This indicates that this marine bacterium does not wet the surface of the PDMS-PU coatings, indicating low bio-adhesion, and is consistent with the PBA results from the screening library. This subset of coatings was also used in an assay with *Ulva linza*, a green algae [48]. The coatings were applied in an array format and the algae allowed to grow on the coatings. Subsequently, the coatings were subjected to a waterjet at different pressures. As can be seen in Figure 1.9, the algae were able to be cleaned from the PDMS-PU coatings at a low waterjet pressure while a much higher pressure was required to remove the algae from the PDMS-PCL-PU coatings. These results are a further indication that the SiPU coatings based on APT-PDMS as the siloxane component have promise as fouling-release coatings.

Hydroxy alkyl carbamate terminated triblock and H-block copolymers of PDMS and PCL were also explored using high-throughput methods to understand the effect of the bulk of the coatings on coating properties and morphology [49]. For this, APT-PDMS was reacted

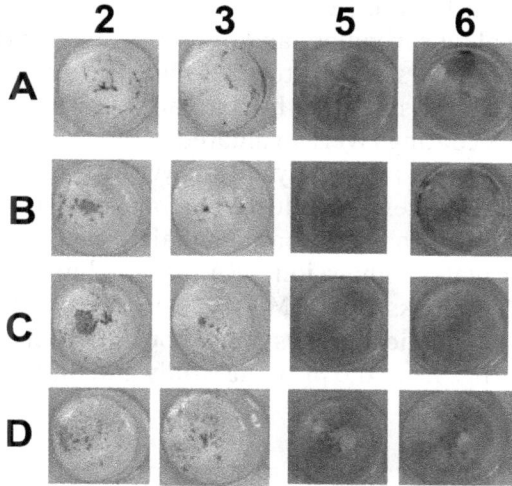

Figure 1.8 Images of PDMS-PU (columns 2-3) and PDMS-PCL-PU (columns 5-6) coatings after drying and CV staining of retained *Cytophaga lytica* biofilms (18 h). Formula designations can be seen in the original article.

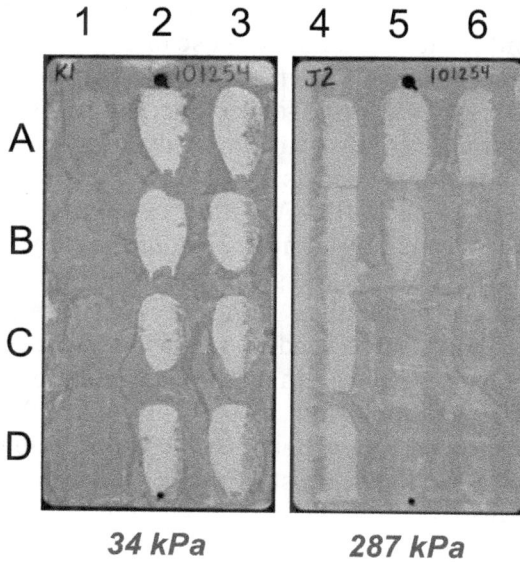

Figure 1.9 Images of coating array panels after settlement of *U. linza* and waterjetting the coatings at the pressures indicated at the bottom of the images. Columns 2-3 are PDMS-PU coatings and columns 5-6 are PDMS-PCL-PU coatings. Columns 1 and 4 are silicone elastomer controls.

with ethylene carbonate (EC) or glycerine carbonate (GC) to form novel hydroxyalkyl carbamate and dihydroxyalkyl carbamate terminated PDMS oligomers. Laboratory FR tests were conducted on SiPU formulations comprising the hydroxyalkyl carbamate-terminated PDMS [50]. The coatings were evaluated using the laboratory biological assays, and were also analyzed for WCA/SE and PBA behavior. The surface analysis tests, conducted before and after 30 days water immersion, showed that the coatings maintained their hydrophobicity even after water immersion. PBA was the lowest for the coating without any PCL blocks and PDMS MW = 15k g/mol. The results with the laboratory biological assays were not as promising as with the PDMS-PU coatings from the previous study and further work was not pursued.

1.4.2 Effect of Acrylic Polyol Composition

These previous experiments were conducted to understand the effect of the PDMS component in the SiPU coatings system. The polyol used in these sets of coatings was a polycaprolactone (PCL) triol. As these coatings are designed for constant water immersion, it was of interest to use a polyol that would be less susceptible to hydrolysis. Thus, a study was conducted to evaluate the use of an acrylic polyol in place of PCL [51]. Acrylic polyols are made by the free radical copolymerization of acrylic acid ester and/or methacrylic acid ester monomers. Hydroxyl functional groups can be easily introduced into the AP chains using monomers such as hydroxyethyl acrylate (HEA) or hydroxyethyl methacrylate (HEMA). Other monomers can be included to vary the glass transition temperature of the polyol. To explore a range of hydroxyl contents and glass transition temperatures (T_g) of the coatings, three monomers, HEA, butyl acrylate (BA) and butyl methacrylate (BMA), were used to prepare a polyol library. The library of 24 polyols was used in an SiPU formulation, along with an APT-PDMS of 10 k g/mol used at 10% in the formulation. Measurement of WCA showed values between 95-100 °C due to self-stratification of the PDMS chains. Most of the coatings showed a low PBA force. Laboratory assays showed that many of the coatings had diatom *N. incerta* release performance comparable to the silicones standards. The removal of *U. linza* showed some interesting trends as a function of acrylic polyol composition, as shown in Figure 1.10. The best removal was achieved across all compositions with the highest amount of hydroxyl monomer, which resulted in higher crosslink density coatings. Also, better removal resulted with the lower T_g co-polymers (higher amount of BA monomer). Past research has shown that FR performance of the coatings is better for coatings with low modulus, which is consistent with this latter trend [8,11]. The results from this experiment revealed that the properties of the coatings can be altered by the composition of the polyol component. Based on these results, an acrylic polyol comprising of BA and HEA was used for future SiPU formulations.

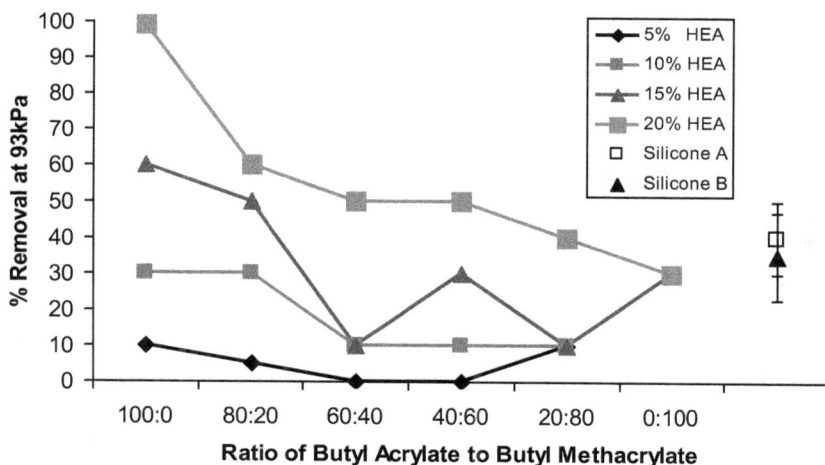

Figure 1.10 Removal of *Ulva linza* sporelings at 93 kPa for the SiPU coatings as a function of acrylic polyol composition. Reproduced from Reference 51 with permission from Federation of Societies for Coatings Technology and Oil and Colour Chemists' Association.

1.4.3 Effect of Polyisocyanate Crosslinker

After tuning the PDMS and the polyol components, a study was conducted to further evaluate the isocyanate crosslinker in the SiPU coatings. Experiments were carried out using polyisocyanates based on hexamethylene diisocyanate (HDI trimer) and isophorone diisocyanate (IPDI trimer) as crosslinkers to generate SiPU formulations with APT-PDMS and acrylic polyol [52]. SE of the coatings was measured initially and periodically after 1, 3 and 7 months of water immersion. Prior to water immersion, the SE of the SiPU coatings with 1-15% APT-PDMS was in the range 15.97-22.38 mN/m. Most of the coatings remained stable and hydrophobic even after 7 months of immersion. As expected, PBA of the coatings decreased with addition of APT-PDMS, indicating the presence of PDMS chains on the surface of the coatings. Formulations containing 20 k and 30 k g/mol analogs showed significantly lower PBA force compared to the other formulations. In general, algal removal increased with increasing APT-PDMS content. The mean average reattachment strength of adult barnacles was lower for IPDI trimer crosslinked coatings than HDI crosslinked coatings. Thus, from these experiments, future work has focused on using IPDI trimer as polyisocyanate crosslinker.

1.4.4 Monofunctional PDMS

The original SiPU systems made use of an α,ω-terminated APT-PDMS as the siloxane component. While more difficult to synthesize, it is possible to make PDMS having a reactive functional group at a single chain end. PDMS anchored only at the chain end in the SiPU system is expected to provide additional degrees of freedom to the surface and possibly yield better FR performance. Thus, a series of APT-PDMS-M with a range of molecular weights were synthesized and incorporated into SiPU coating formulation [53]. The coatings exhibited high WCA and low SE values indicating formation of hydrophobic coatings. The higher MW APT-PDMS-M showed less change in WCA after water immersion. The coatings, irrespective of the MW and amount of the PDMS macromers showed low PBA force. Overall, the experimental coatings showed better FR performance against biofilm than a PU control and silicone standards. Against diatoms (*N. incerta*), most of the formulations with 10% APT-PDMS-M deterred their attachment. The removal decreased as MW of APT-PDMS-M increased. Some low MW APT-PDMS-M formulations showed higher diatom release as compared to PU control, which may be attributed to increase in SE after water immersion. Conversely, higher *U. linza* removal was observed on coatings with higher MW APT-PDMS-M, with FR performance similar to the commercial standards. Against barnacles, high MW APT-PDMS-M facilitated higher removal with low adhesion forces, with the formulations having 5 k (10%), 10 k (5, 10%) and 15 k (5, 10%) APT-PDMS-M showing the lowest adhesion force.

1.5 Laboratory vs. Field Testing Study

Based on the previous findings, it was decided to carry out a matrix experiment using several of the best performing compositions and study these thoroughly using the laboratory methods as well as subject the coatings to field testing at three different field testing sites around the world [54]. The compositions of the eight SiPU coatings are shown in Table 1.1. APT-PDMS of molecular weight 30,000 g/mol was used in all cases at either 10 or 20 wt%. Both monofunctional (M) and difunctional (D) PDMS were used in the experiment. The coatings also contained either polycaprolactone or acrylic polyol. Commercial FR coatings Intersleek (IS) 700 and 900 were used for comparison.

Table 1.1 Compositions of SiPU coatings used in laboratory and field testing study

Sample	PDMS weight %	PDMS functionality	Polyol
PCL-M-10	10	M	Polycaprolactone
PCL-M-20	20	M	Polycaprolactone
PCL-D-10	10	D	Polycaprolactone
PCL-D-20	20	D	Polycaprolactone
ACR-M-10	10	M	Acrylic
ACR-M-20	20	M	Acrylic
ACR-D-10	10	D	Acrylic
ACR-D-20	20	M	Acrylic

In laboratory measurements, the SiPU coatings had WCA values above 100°, indicating that the surfaces were hydrophobic. PBA values were below 10 N for most of the coatings, including the FR controls. Two coatings, PCL-D-20 and ACR-D-20, had PBA values a bit higher, around 20 N. In laboratory tests with marine bacteria *C. lytica*, surface coverage was lower and removal was better for the coatings based on the monofunctional PDMS. The results for the coatings based on the monofunctional PDMS were similar to that of IS 900, while the SiPU coatings made using the difunctional PDMS had similar performance to IS 700. Removal of the diatom *N. incerta* was moderate for difunctional PDMS based coatings, however, was better for the coatings based on monofunctional PDMS. Removal of the alga *U. linza* was significantly better for the coatings generated suing monofunctional PDMS as compared to difunctional PDMS and was comparable to IS 900. Reattached barnacle adhesion was a bit variable, however, PCL-M10 and PCL-M20 coatings were observed to have the lowest values.

This set of SiPU coatings were deployed at three different field test sites: Morro Bay, California (California Polytechnic State University, San Luis Obispo), Indian River Lagoon, Florida (Florida Institute of Technology) and Singapore (National University of Singapore). Over a period of many months, measurements of fouling removal using a water jet and/or barnacle adhesion measurements were taken. In general, the SiPU coatings made using monofucntional PDMS per- formed better than difunctional PDMS. Further, the SiPU coatings based on monofunctional PDMS had similar barnacle adhesion as the Intersleek 700 control coating. Using statistical analysis, correlations were identified between the laboratory assays and the field testing results. An outcome of the analysis indicated that the reattached bar- nacle measurements as well as *C. lytica* retraction assays were the most effective for performance prediction in the field [54].

1.6 SiPU Coating Optimization

Based on the results from the previous study, it was of interest to de- termine if further optimization could be carried out to improve the performance of the SiPU coating system based on difunctional PDMS, as difunctional PDMS is much more easily synthesized and more eco- nomical to produce than monofunctional PDMS. In this study, SiPU coating formulations were prepared using APT-PDMS (20 k g/mol) at 20% and IPDI-based polyisocyanate. Variables studied were the type of polyol (based on PCL or acrylic (AP)) and the solvents used (tolu- ene, MAK or EEP) [55].

Characterization of the cured coatings showed PBA values be- tween 8-20 N and SE between 22-24 mN/m, indicating self-stratifica- tion of PDMS chains. Results for laboratory FR assays with *U. linza* sporelings showed removal was better for the experimental SiPU coatings than the standard silicone coatings (Intersleek 900 (I900), Intersleek 425 (I425) and Silastic T2 9T2)). Experiments also showed that *U. linza* attachment was correlated to the roughness of the coat- ings (determined using optical profilometry); a coating having sur- face roughness of 60 nm showed good *Ulva* removal at 89 kPa, while a coating with roughness of 160 nm had poor *Ulva* removal at 89 kPa. The experimental coatings showed lower *C. lytica* biofilm coverage and better removal than the commercial standards. Between the two polyols, coatings with AP showed better biofilm removal compared to PCL. The roughness of PCL based coatings was generally higher than the AP coatings. With *N. incerta*, coatings made without EEP showed better removal as compared to coatings with higher EEP amounts. Based on the initial results, four additional coatings were made using 20k and 30k APT-PDMS at 10 and 20% loading levels using the best solvent composition identified earlier. A best SiPU coating designated "A4" was identified that had optimum FR in the laboratory assays.

1.7 Surface Analysis of SiPU Coatings

To further understand the nature of the stratification of PDMS to the surface of the SiPU coatings, efforts were made to understand the exact chemical nature of the coating surfaces formed after curing and correlate their FR performance to the type of surface formed. Surface characterization experiments were carried out using X-ray photoelectron spectroscopy (XPS), transmission electron microscopy (TEM), Rutherford backscattering spectroscopy (RBS) and nanoindentation [56]. Herein, we highlight the XPS and RBS studies in detail. For the study, the SiPU coatings described in Table 1.1, along with two coatings from the optimization study, designated A4 and D3, were used [55]. The XPS measurements were made at different take-off angles (10°, 35° and 80°) which correlate to sampling depths of 2.8, 9.4 and 16 nm, respectively. The elemental concentrations were transformed into the amount of PDMS on the surface of the coatings, and is shown in Table 1.2. From the results, it can be clearly observed

Table 1.2 Calculated PDMS content of the coatings at different take-off angles from XPS (data from Reference 56).

Sample	wt% PDMS			Sample	wt% PDMS		
	10°	35°	80°		10°	35°	80°
ACR-M-20	98.7	88.2	77.2	PCL-M-20	87.2	88.2	73.9
ACR-D-20	44.2	31.0	21.0	PCL-D-20	34.2	32.4	22.9
ACR-M-10	89.6	76.7	65.1	PCL-M-10	85.5	79.7	74.9
ACR-D-10	45.7	41.2	28.0	PCL-D-10	35.5	26.9	32.6
A4	52.3	43.3	27.2	D3	93.9	83.2	59.9

that PDMS was enriched in the surface in all cases in excess of the bulk amounts. The amount of PDMS on the surface was also significantly higher for the coatings made using monofunctional PDMS than difunctional PDMS. PDMS anchored at one chain end into the coating matrix had greater degrees of freedom than PDMS anchored at both chain ends, resulting in higher PDMS content on the surface. Coatings made using acrylic polyol also had a slightly higher PDMS surface content than the coatings made using PCL as polyol. An exception was coating D3, which had essentially the same composition as PCL-D-20 but a different solvent system, indicating that changing the solvent composition can significantly affect the stratification of PDMS.

RBS is a non-destructive and multi-elemental analysis technique, which gives elemental depth profiles with a depth resolution of 5-50 nm and a maximum depth of 2-20 μm. Samples ACR-M-20 and ACR-D-20 were analyzed using RBS. It was found that a 7 nm thick PDMS layer was present in ACR-M-20, while for ACR-D-20, the surface PDMS layer was 3.5 nm thick. This confirmed the XPS results that the SiPU coating made with monofunctional PDMS had a significantly greater amount of PDMS on the surface than difunctional PDMS.

Further XPS experiments were carried out on formulation "A4" using an argon ion etching technique to determine the change in composition as a function of depth in the coating [57]. As illustrated in Figure 1.11, it can be seen that the signal for silicon (from PDMS) is highest at the outermost surface of the coating, but decays rapidly as the surface is etched away, further illustrating the stratification in the SiPU coatings.

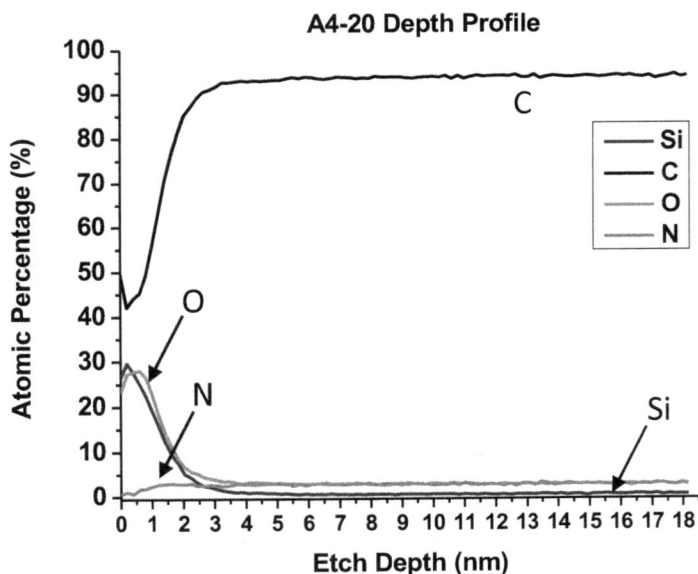

Figure 1.11 Atomic concentrations determined by XPS as a function of etching depth into the coating. Reproduced from Reference 57 with permission from American Chemical Society.

1.8 Modifications of SiPU Formulations

1.8.1 Effect of Pigmentation

Pigments are typically incorporated into marine coatings to impart color. While the presence of color imparts a pleasing appearance, it also helps the coating applicator see where the coating has been applied in order to be able to apply a uniform coating to the ship hull. Therefore, a study was carried out with the addition of a simple white pigment, titanium dioxide (TiO_2) R-706, to the SiPU coating system [58]. The primary aim of the study was to determine if the presence of the pigment affected the fouling-release performance of the coating. Formulations were made at 0, 20 and 30 pigment volume concentration (PVC) and 0, 10, 20 and 30 wt% APT-PDMS (~27 k g/mol). The pigment was dispersed into acrylic polyol. The SiPU formulations consisted of the pigment grinds, IPDI-based polyisocyanate, APT-PDMS, DBTDAc catalyst, solvents and additives. Overall, the surface properties of the coatings were not significantly affected by the presence of the pigment. The WCA for PDMS containing coatings were greater than 100°, and surface energy values in the range expected for PDMS. For the laboratory biological assays, removal of *C. lytica* biofilm was slightly lower for the pigmented SiPU coatings compared to the non-pigmented control, but still in the range of the silicone standards. Removal of barnacles was similar for the pigmented coatings compared to the non-pigmented coatings. An interesting result was that the removal of the diatom *N. incerta* seemed to be improved due to the presence of the pigment.

A subsequent experiment further explored the use of TiO_2 pigment in SiPU coatings using monofunctional PDMS [59]. As with the SiPU coatings made using difunctional PDMS, the incorporation of pigment had a little effect on the surface energy and fouling-release properties of the coatings, analyzed using laboratory biological assays.

1.8.2 Incorporation of Silicone Oils

Silicone oils are commonly added into silicone elastomer fouling-release coatings, and this concept was initially patented by Milne [60]. It is believed that the silicone oils can migrate to the surface of the coating to provide additional interfacial slippage. To adapt this concept to the SiPU coatings system, an initial study was carried out where PDMS oils were added to SiPU formulations [52]. Moreover, four APT-PDMS in different amounts were used as reactive PDMS component in the SiPU coatings. The APT-PDMS component in the SiPU formulations is expected to work as a reservoir for the free silicone oil, which will migrate to the surface over the lifetime of the coatings. As expected, the coatings displayed hydrophobic character, however, the inclusion of silicone oil actually resulted in poorer fouling-release performance characterized using laboratory biological assays. It is hypothesized that the PDMS oils were "hidden" inside the PDMS, internal to the SiPU coating, thus, reducing the amount of PDMS present on the surface.

As a follow up to the previous methyl silicone oil-modified study, SiPU formulations were developed by adding homopolymer or copolymer phenylmethyl silicone oils [57]. Due to the slight incompatibility of phenyl-containing silicone oils with PDMS, it was expected that the use of these oils would increase surface stratification of the silicone oils to form highly lubricated coatings. The oils were added to the SiPU coating formulations at 1, 2 and 5 wt%. In general, WCA of the coatings showed hydrophobic nature with WCA values between 90-105°. The coatings maintained their hydrophobicity after 28 days of water immersion. Depth profiling of the formulations using XPS showed that the concentration of Si decreased from 22% at the surface to 5% ~3.5 nm into the bulk of the coating, indicating that silicone chains remain concentrated at the top 3.5 nm layer of the coating surface, similar to results obtained by Siripirom [56]. In laboratory biological assays, the incorporation of the silicone oils did not affect the removal of the bacteria *C. lytica* or diatom *N. incerta*. For the green alga *U. linza*, removal was better for the oil-modified SiPU coatings than the control SiPU coating and silicone controls, including Intersleek 900. A number of oil-modified coatings showed good performance in the barnacle reattachment assay, with low barnacle adhesion strength. Mussel adhesion assays also identified coatings with either no attachment at all or very low adhesion. Selected coatings were evaluated in field immersion testing in Hawaii, California and Singapore. It was demonstrated that the incorporation of silicone oil yielded better fouling-release than the SiPU coatings without oil.

1.9 Amphiphilic SiPU Coatings

Due to different (and not very well understood) adhesion mechanisms of thousands of different marine fouling species, it has been difficult to develop a universal coating which can successfully deter the attachment of a broad spectrum of fouling organisms. For example, while silicone-based coatings are effective for barnacle and green algal fouling, diatom algae and some marine bacteria adhere very well to silicone surfaces, creating a brown or black slime. One approach to design a coating that resists adhesion by a broader range of fouling organisms is the so-called amphiphilic coating system. The hypothesis of this approach is that a coating having areas of mixed surface energy in the form of nanodomains presents an "ambiguous" surface for fouling organisms, reducing the ability of the various bioadhesives to wet and spread on the surface. A number of approaches have been explored in the literature that involve synthetic strategies to combine hydrophobic chain segments, usually based on fluoropolymers, siloxanes or hydrocarbons, with hydrophilic chains, most typically poly(ethyleneglycol)s [61-84]. In the commercial arena, Akzo-Nobel International Paint introduced Intersleek 900, which is described as an amphiphilic fluoropolymer fouling-release coating, followed by Intersleek 1100SR, an amphiphilic coating with better slime removal properties. However, the specific chemical compositions of these coatings is not clear.

Due to the success of amphiphilic coating approaches, it was of interest to explore this concept in the self-stratified SiPU coatings. In the SiPU system, in order to ensure that both the hydrophobic and hydrophilic components are expressed on the surface, the hydrophilic component is chemically bound to the hydrophobic component (PDMS). PDMS will stratify to the surface due to its low surface energy and bring the chemically bound hydrophilic component along with it. Studies have been carried out using hydrophilic components such as carboxylic acids, poly(ethylene glycol)s (PEGs) and zwitterionic compounds.

In one study, carboxylic acid groups were attached to the PDMS component to yield amphiphilic SiPU coatings [85]. Another approach involved the synthesis of siloxane-zwitterionic triblock copolymers that were incorporated into the SiPU system [86]. With both of these systems, laboratory biological assays indicated changes in surface composition and organism adhesion, however, none of these approaches resulted in an optimum fouling-release performance for multiple marine organisms.

In a preliminary study, APT-PDMS having pendant PEG chains was developed and used in the SiPU coating system [52,87]. Coatings compositions having good release properties for multiple organisms, such as *U. linza* and *N. incerta*, were identified.

Galhenage *et al.* [88] explored amphiphilic coatings using the pre-polymer approach by reacting monofunctional PEG and PDMS of different MW with IPDI trimers, which were then incorporated into the SiPU formulation. The amphiphilic coatings showed similar trends with respect to WCA values; the values decreased after water immersion, indicating rearrangement of chains on the surface. The amphiphilic coatings showed good FR performance against *U. linza*, with a number of coatings also exhibiting good release of *N. incerta*. In case of hard foulants, several formulations required very low force for removal for barnacles, and most formulations could deter mussel attachment completely. Therefore, the results showed that hydrophilic moieties can be easily incorporated into SiPU coatings to impart hydrophilicity to the coatings. Interestingly, some of these coatings also showed potential as low ice-adhesion coatings, thus confirming their versatility [89].

1.10 Conclusions

Hybrid self-stratified SiPU coatings show great potential for use in practical marine applications. Siloxane chains migrate to the coating-air interface upon crosslinking, thus, providing a hydrophobic character to the coatings. The formulations remain unaltered even after months of water immersion due to the highly crosslinked nature of the coatings. Monofunctional PDMS chains tend to self-stratify to a higher extent as compared to the difunctional components, however, solvent composition also plays a significant role in determining the extent of stratification. SiPU coating formulations have been demonstrated in the field to have comparable performance to early-generation silicone elastomer coatings. It has also been demonstrated that the amphiphilic SiPU coatings can be obtained using various synthetic techniques, with many coatings showing broader-spectrum fouling release characteristics. Additional developments are underway to further improve the performance of the coatings for a broad range of marine organisms.

Acknowledgements

The support of the Office of Naval Research for the research reported in this chapter is gratefully acknowledged by the authors. The writing of this chapter is supported under ONR grant number N00014-16-1-3064.

References

1. Yebra, D. M., Kiil, S., and Dam-Johansen, K. (2004) Antifouling technology-past, present and future steps towards efficient and environmentally friendly antifouling coatings. *Progress in Organic Coatings*, **50**(2), 75-104.
2. Schultz, M. P., Bendick, J. A., Holm, E. R., and Hertel, W. M. (2011) Economic impact of biofouling on a naval surface ship. *Biofouling*, **27**(1), 87-98.
3. Callow, J. A., and Callow, M. E. (2011) Trends in the development of environmentally friendly fouling-resistant marine coatings. *Nature Communications*, **2**, 244.
4. Lejars, M., Margaillan, A., and Bressy, C. (2012) Fouling release coatings: a nontoxic alternative to biocidal antifouling coatings. *Chemical Reviews*, **112**(8), 4347-4390.
5. Ciriminna, R., Bright, F. V., and Pagliaro, M. (2015) Ecofriendly antifouling marine coatings. *ACS Sustainable Chemistry & Engineering*, **3**(4), 559-565.
6. Champ, M. A. (2000) A review of organotin regulatory strategies, pending actions, related costs and benefits. *Science of the Total Environment*, **258**, 21-71.
7. Magin, C. M., Cooper, S. P., and Brennan, A. B. (2010) Non-toxic antifouling strategies. *Materials Today*, **13**(4), 36-44.
8. Brady, R. F. (1999) Properties which influence marine fouling resistance in polymers containing silicon and fluorine. *Progress in Organic Coatings*, **35**(1-4), 31-35.
9. Brady, R. F. (2001) A fracture mechanical analysis of fouling release from nontoxic antifouling coatings. *Progress in Organic Coatings*, **43**(1-3), 188-192.
10. Brady, R. F. (2000) Clean hulls without poisons: Devising and testing nontoxic marine coatings. *Journal of Coatings Technology*, **72**(900), 45-56.
11. Brady, Jr., R. F., and Singer, I. L. (2000) Mechanical factors favoring release from fouling release coatings. *Biofouling*, **15**(1-3), 73-81.
12. Briganti, E., Losi, P., Raffi, A., Scoccianti, M., Munao, A., and Soldani, G. (2006) Silicone based polyurethane materials: a promising biocompatible elastomeric formulation for cardiovascular applications. *Journal of Materials Science: Materials in Medicine*, **17**(3), 259-266.
13. Chen, J., and Gardella, Jr., J. A. (1998) Solvent effects on the surface composition of poly(dimethylsiloxane)-co-polystyrene/polystyrene blends. *Macromolecules*, **31**(26), 9328-9336.
14. Chen, X., Gardella, Jr., J. A., Ho, T., and Wynne, K. J. (1995) Surface composition of a series of dimethylsiloxane urea urethane segmented copolymers studied by electron spectroscopy for chemical

analysis. *Macromolecules*, **28**(5), 1635-1642.

15. Gardella, Jr., J. A., Ho, T., Wynne, K. J., and Zhuang, H.-Z. (1995) Using solubility difference to achieve surface phase separation in dimethylsiloxane-urea-urethane copolymers. *Journal of Colloid and Interface Science*, **176**(1), 277-279.

16. Hwang, S. S., Ober, C. K., Perutz, S., Iyengar, D. R., Schneggenburger, L. A., and Kramer, E. J. (1995) Block copolymers with low surface energy segments: siloxane- and perfluoroalkane-modified blocks. *Polymer*, **36**(6), 1321-1325.

17. Okkema, A. Z., Fabrizius, D. J., Grasel, T. G., Cooper, S. L., and Zdrahala, R. J. (1989) Bulk, surface and blood-contacting properties of polyether polyurethanes modified with polydimethylsiloxane macroglycols. *Biomaterials*, **10**(1), 23-32.

18. Yilgor, I., and McGrath, J. E. (1988) Polysiloxane-containing copolymers: A survey of recent developments. *Advances in Polymer Science*, **86**, 1-86.

19. Patel, N. M., Dwight, D. W., Hedrick, J. L., Webster, D. C., and McGrath, J. E. (1988) Surface and bulk phase separation in block copolymers and their blends. Polysulfone/polysiloxane. *Macromolecules*, **21**(9), 2689-2696.

20. Ho, T., Wynne, K. J., and Nissan, R. A. (1993) Polydimethylsiloxane-urea-urethane copolymers with 1,4-benzenedimethanol as chain extender. *Macromolecules*, **26**(25), 7029-7036.

21. Wynne, K. J., Ho, T., Nissan, R. A., Chen, X., and Gardella, Jr., J. A.. (1994) Poly(dimethylsiloxane)-urea-urethane copolymers. Synthesis and surface properties. *ACS Symposium Series*, **572**, 64-80.

22. Pike, J. K., Ho, T., and Wynne, K. J. (1996) Water-induced surface rearrangements of poly(dimethylsiloxane-urea-Urethane) segmented block copolymers. *Chemistry of Materials*, **8**(4), 856-860.

23. Tezuka, Y., Kazama, H., and Imai, K. (1991) Environmentally induced macromolecular rearrangement on the surface of polyurethane-polysiloxane block copolymers. *Journal of the Chemical Society, Faraday Transactions*, **87**(1), 147-152.

24. Chisholm, B. J., Christianson, D., and Webster, D. C. (2006) Combinatorial materials research applied to the development of new surface coatings - II. Process capability analysis of the coating formulation workflow. *Progress in Organic Coatings*, **57**(2), 115-122.

25. Chisholm, B., Potyrailo, R., Cawse, J., Brennan, M., Molaison, C., Shaffer, R., Whisenhunt, D., and Olson, D. (2002) Combinatorial Chemistry Methods for Coating Development III. An Illustration of an Experiment Conducted with the Combinatorial Factory. *Proceedings of the International Waterborne, High-Solids, and Powder Coatings Symposium*, pp. 125-137.

26. Chisholm, B., Potyrailo, R., Cawse, J., Shaffer, R., Brennan, M., Molaison, C., Whisenhunt, D., Flanagan, B., Olson, D., Akhave, J., Saund-

ers, D., Mehrabi, A., and Licon, M. (2002) The development of combinatorial chemistry methods for coating development I. Overview of the experimental factory. *Progress in Organic Coatings*, **45**(2-3), 313-321.

27. Chisholm, B., Potyrailo, R., Shaffer, R., Cawse, J., Brennan, M., and Molaison, C. (2003) Combinatorial chemistry methods for coating development: III. Development of a high throughput screening method for abrasion resistance: correlation with conventional methods and the effects of abrasion mechanism. *Progress in Organic Coatings*, **47**(2), 112-119.

28. Chisholm, B., Potyrailo, R., Shaffer, R., Cawse, J., Brennan, M., and Molaison, C. (2003) Combinatorial chemistry methods for coating development VI. Correlation of high throughput screening methods with conventional measurement techniques. *Progress in Organic Coatings*, **48**(2-4), 219-226.

29. Chisholm, B., Stafslien, S., Christianson, D., Gallagher-Lein, C., Daniels, J., Rafferty, C., Wal, L. V., and Webster, D. C. (2007) Combinatorial materials research applied to the development of new surface coatings - VIII: Overview of the high-throughput measurement systems developed for a marine coating workflow. *Applied Surface Science*, **254**(3), 692-698.

30. Chisholm, B. J., and Webster, D. C. (2007) The development of coatings using combinatorial/high throughput methods: a review of the current status. *Journal of Coatings Technology and Research*, **4**(1), 1-12.

31. Chisholm, B., Webster, D. C., Bennett, J., Berry, M., Christianson, D., Kim, J., Mayo, B., and Gubbins, N. (2007) Combinatorial materials research applied to the development of new surface coatings VII: An automated system for adhesion testing. *Review of Scientific Instruments*, **78**(7), 072213.

32. Chisholm, B. J., Potyrailo, R. A., Cawse, J. N., Shaffer, R. E., Brennan, M., and Molaison, C. A. (2003) Combinatorial chemistry methods for coating development: IV. The importance of understanding process capability. *Progress in Organic Coatings*, **47**(2), 120-127.

33. Chisholm, B. J., Stafslien, S. J., Majumdar, P., and Webster, D. (2008) Fast-moving solutions: automated processes speed up development of marine antifouling coatings. *European Coatings Journal*, **11**, 32.

34. He, J., Bahr, J., Chisholm, B. J., Li, J., Chen, Z., Balbyshev, S. N., Bonitz, V., and Bierwagen, G. P. (2008) Combinatorial materials research applied to the development of new surface coatings X: A high-throughput electrochemical impedance spectroscopy method for screening organic coatings for corrosion inhibition. *Journal of Combinatorial Chemistry*, **10**(5), 704-713.

35. Webster, D. C., Chisholm, B. J., and Stafslien, S. J. (2007) Mini-review:

Combinatorial approaches for the design of novel coating systems. *Biofouling*, **23**(3/4), 179-192.

36. Webster, D. C., Chisholm, B. J., and Stafslien, S. J. (2009) High throughput methods for the design of fouling control coatings. In: *Advances in Marine Antifouling Coatings and Technologies*, Hellio, C., and Yebra, D. (eds.). Woodhead Publishing, UK, pp. 365-392.

37. Majumdar, P., Ekin, A., and Webster, D. C. (2007) Thermoset siloxane-urethane fouling release coatings. *ACS Symposium Series*, **957**, 61-75.

38. Majumdar, P., and Webster, D. C. (2005) Preparation of siloxane–urethane coatings having spontaneously formed stable biphasic microtopograpical surfaces. *Macromolecules*, **38**(14), 5857-5859.

39. Majumdar, P., and Webster, D. C. (2006) Influence of solvent composition and degree of reaction on the formation of surface microtopography in a thermoset siloxane–urethane system. *Polymer*, **47**(11), 4172-4181.

40. Majumdar, P., Stafslien, S., Daniels, J., and Webster, D. C. (2007) High throughput combinatorial characterization of thermosetting siloxane–urethane coatings having spontaneously formed microtopographical surfaces. *Journal of Coatings Technology and Research*, **4**(2), 131-138.

41. Rittschof, D., Orihuela, B., Stafslien, S., Daniels, J., Christianson, D., Chisholm, B., and Holm, E. (2008) Barnacle reattachment: a tool for studying barnacle adhesion. *Biofouling*, **24**(1), 1-9.

42. Majumdar, P., and Webster, D. C. (2007) Surface microtopography in siloxane–polyurethane thermosets: The influence of siloxane and extent of reaction. *Polymer*, **48**(26), 7499-7509.

43. Webster, D. C. Unpublished results.

44. Stafslien, S. J., Bahr, J., Daniels, J., Christianson, D. A., and Chisholm, B. J. (2011) High-throughput screening of fouling-release properties: an overview. *Journal of Adhesion Science and Technology*, **25**(17), 2239-2253.

45. Ekin, A., and Webster, D. C. (2006) Library synthesis and characterization of 3-aminopropyl-terminated poly (dimethylsiloxane) s and poly (ε-caprolactone)-b-poly (dimethylsiloxane) s. *Journal of Polymer Science Part A: Polymer Chemistry*, **44**(16), 4880-4894.

46. Ekin, A., and Webster, D. C. (2007) Combinatorial and high-throughput screening of the effect of siloxane composition on the surface properties of crosslinked siloxane-polyurethane coatings. *Journal of Combinatorial Chemistry*, **9**(1), 178-188.

47. Stafslien, S., Daniels, J., Mayo, B., Christianson, D. A., Chisholm, B., Ekin, A., Webster, D. C., and Swain, G. W. (2007) Combinatorial materials research applied to the development of new surface coatings. IV: A high-throughput bacterial retention and retraction assay for screening fouling-release performance of coatings. *Biofouling*, **23**,

45-54.

48. Casse, F., Ribeiro, E., Ekin, A., Webster, D. C., Callow, J. A., and Callow, M. E. (2007) Laboratory screening of coating libraries for algal adhesion. *Biofouling*, **23**(4), 267-276.

49. Ekin, A., and Webster, D. C. (2006) Synthesis and characterization of novel hydroxyalkyl carbamate and dihydroxyalkyl carbamate terminated poly (dimethylsiloxane) oligomers and their block copolymers with poly (ε-caprolactone). *Macromolecules*, **39**(25), 8659-8668.

50. Ekin, A., Webster, D. C., Daniels, J. W., Stafslien, S. J., Cassé, F., Callow, J. A., and Callow, M. E. (2007) Synthesis, formulation, and characterization of siloxane–polyurethane coatings for underwater marine applications using combinatorial high-throughput experimentation. *Journal of Coatings Technology and Research*, **4**(4), 435-451.

51. Pieper, R. J., Ekin, A., Webster, D. C., Casse, F., Callow, J. A., and Callow, M. E. (2007) Combinatorial approach to study the effect of acrylic polyol composition on the properties of crosslinked siloxane-polyurethane fouling-release coatings. *Journal of Coatings Technology and Research*, **4**(4), 453-461.

52. Bodkhe, R. B. (2011) *Amphiphilic Siloxane-Polyurethane Coatings*, Ph.D. Thesis, North Dakota State University, USA.

53. Sommer, S., Ekin, A., Webster, D. C., Stafslien, S. J., Daniels, J., VanderWal, L. J., Thompson, S. E. M., Callow, M. E., and Callow, J. A. (2010) A preliminary study on the properties and fouling-release performance of siloxane–polyurethane coatings prepared from poly (dimethylsiloxane)(PDMS) macromers. *Biofouling*, **26**(8), 961-972.

54. Stafslien, S. J., Sommer, S., Webster, D. C., Bodkhe, R., Pieper, R., Daniels, J., Vander Wal, L., Callow, M. C., Callow, J. A., Ralston, E., Swain, G., Brewer, L., Wendt, D., Dickinson, G. H., Lim, C.-S., and Teo, S. L.-M. (2016) Comparison of laboratory and field testing performance evaluations of siloxane-polyurethane fouling-release marine coatings. *Biofouling*, **32**(8), 949-968.

55. Bodkhe, R. B., Thompson, S. E. M., Yehle, C., Cilz, N., Daniels, J., Stafslien, S. J., Callow, M. E., Callow, J. A., and Webster, D. C. (2012) The effect of formulation variables on fouling-release performance of stratified siloxane–polyurethane coatings. *Journal of Coatings Technology and Research*, **9**(3), 235-249.

56. Siripirom, C. (2012) High-throughput *Methods for Characterizing the Mechanical Properties of Coatings*, Ph.D. Thesis, North Dakota State University, USA.

57. Galhenage, T. P., Hoffman, D., Silbert, S. D., Stafslien, S. J., Daniels, J., Miljkovic, T., Finlay, J. A., Franco, S. C., Clare, A. S., Nedved, B. T., Hadfield, M. G., Wendt, D. E., Waltz, G., Brewer, L., Teo, S. L. M., Lim, C.-S., and Webster, D. C. (2016) Fouling-release performance of silicone

oil-modified siloxane-polyurethane coatings. *ACS Applied Materials and Interfaces*, **8**(42), 29025-29036.

58. Sommer, S. A., Byrom, J. R., Fischer, H. D., Bodkhe, R. B., Stafslien, S. J., Daniels, J., Yehle, C., and Webster, D. C. (2011) Effects of pigmentation on siloxane–polyurethane coatings and their performance as fouling-release marine coatings. *Journal of Coatings Technology and Research*, **8**(6), 661-670.

59. Sommer, S. A. (2011) *Siloxane-Polyurethane Fouling-release Coatings based on PDMS Macromers*, Ph.D. Thesis, North Dakota State University, USA.

60. Milne, A. (1977) Anti-fouling Marine Compositions, patent US4025693.

61. Gudipati, C. S., Finlay, J. A., Callow, J. A., Callow, M. E., and Wooley, K. L. (2005) The antifouling and fouling-release performance of hyperbranched fluoropolymer (HBFP)-poly(ethylene glycol) (PEG) composite coatings evaluated by adsorption of biomacromolecules and the green fouling alga Ulva. *Langmuir*, **21**(7), 3044-3053.

62. Gudipati, C. S., Greenlief, C. M., Johnson, J. A., Prayongpan, P., and Wooley, K. L. (2004) Hyperbranched fluoropolymer and linear poly(ethylene glycol) based amphiphilic crosslinked networks as efficient antifouling coatings: An insight into the surface compositions, topographies, and morphologies. *Journal of Polymer Science, Part A: Polymer Chemistry*, **42**(24), 6193-6208.

63. Pollack, K. A., Imbesi, P. M., Raymond, J. E., and Wooley, K. L. (2014) Hyperbranched fluoropolymer-polydimethylsiloxane-poly(ethylene glycol) cross-linked terpolymer networks designed for marine and biomedical applications: heterogeneous nontoxic antibiofouling surfaces. *ACS Applied Materials and Interfaces*, **6**(21), 19265-19274.

64. Bartels, J. W., Imbesi, P. M., Finlay, J. A., Fidge, C., Ma, J., Seppala, J. E., Nystrom, A. M., Mackay M. E., Callow, J. A., Callow, M. E., and Wooley, K. L. (2011) Antibiofouling hybrid dendritic Boltorn/star PEG thiolene cross-linked networks. *ACS Applied Materials and Interfaces*, **3**(6), 2118-2129.

65. Felder, S. E., Imbesi, P. M., Fidge, C., and Wooley, K. L. (2011) Tunable, amphiphilic hyperbranched fluoropolymers (HBFPs) as antibiofouling coatings: an investigation of the molecular- to microscale tunability. *Polymer Preprints (American Chemical Society)*, **52**(1).

66. Krishnan, S., Ayothi, R., Hexemer, A., Finlay, J. A., Sohn, K. E., Perry, R., Ober, C. K., Kramer, E. J., Callow, M. E., Callow, J. A., and Fischer, D. A. (2006) Anti-biofouling properties of comblike block copolymers with amphiphilic side chains. *Langmuir*, **22**(11), 5075-5086.

67. Krishnan, S., Wang, N., Ober, C. K., Finlay, J. A., Callow, M. E., Callow, J. A., Hexemer, A., Sohn, K. E., Kramer, E. J., and Fischer, D. A. (2006)

Comparison of the fouling release properties of hydrophobic fluorinated and hydrophilic PEGylated block copolymer surfaces: attachment strength of the diatom Navicula and the green alga Ulva. *Biomacromolecules*, **7**(5), 1449-1462.

68. Krishnan, S., Weinman, C. J., and Ober, C. K. (2008) Advances in polymers for anti-biofouling surfaces. *Journal of Materials Chemistry*, **18**(29), 3405-3413.

69. Andruzzi, L., Chiellini, E., Galli, G., Li, X., Kang, S. H., and Ober, C. K. (2002) Engineering low surface energy polymers through molecular design: synthetic routes to fluorinated polystyrene-based block copolymers. *Journal of Materials Chemistry*, **12**(6), 1684-1692.

70. Cho, Y., Cho, D., Park, J. H., Frey, M. W., Ober, C. K., and Joo, Y. L. (2012) Preparation and characterization of amphiphilic triblock terpolymer-based nanofibers as antifouling biomaterials. *Biomacromolecules*, **13**(5), 1606-1614.

71. Cho, Y., Sundaram, H. S., Weinman, C. J., Paik, M. Y., Dimitriou, M. D., Finlay, J. A., Callow, M. E., Callow, J. A., Kramer, E. J., and Ober, C. K. (2011) Triblock copolymers with grafted fluorine-free, amphiphilic, non-ionic side chains for antifouling and fouling-release applications. *Macromolecules*, **44**(12), 4783-4792.

72. Weinman, C. J., Finlay, J. A., Park, D., Paik, M. Y., Krishnan, S., Sundaram, H. S., Dimitriou, M., Sohn, K. E., Callow, M. E., Callow, J. A., Handlin, D. L., Willis, C. L., Kramer, E. J., and Ober, C. K. (2009) ABC triblock surface active block copolymer with grafted ethoxylated fluoroalkyl amphiphilic side chains for marine antifouling/fouling-release applications. *Langmuir*, **25**(20), 12266-12274.

73. Youngblood, J. P., Andruzzi, L., Ober, C. K., Hexemer, A., Kramer, E. J., Callow, J. A., Finlay, J. A., and Callow, M. E. (2003) Coatings based on side-chain ether-linked poly(ethylene glycol) and fluorocarbon polymers for the control of marine biofouling. *Biofouling*, **19**, 91-98.

74. Zhou, Z., Calabrese, D. R., Taylor, W., Finlay, J. A., Callow, M. E., Callow, J. A., Fischer, D., Kramer, E. J., and Ober, C. K. (2014) Amphiphilic triblock copolymers with PEGylated hydrocarbon structures as environmentally friendly marine antifouling and fouling-release coatings. *Biofouling*, **30**(5), 589-604.

75. Park, D., Weinman, C. J., Finlay, J. A., Fletcher, B. R., Paik, M. Y., Sundaram, H. S., Dimitriou, M. D., Sohn, K. E., Callow, M. E., Callow, J. A., Handlin, D. L., Willis, C. L., Fischer, D. A., Kramer, E. J., and Ober, C. K. (2010) Amphiphilic surface active triblock copolymers with mixed hydrophobic and hydrophilic side chains for tuned marine fouling-release properties. *Langmuir*, **26**(12), 9772-9781.

76. Marabotti, I., Morelli, A., Orsini, L. M., Martinelli, E., Galli, G., Chiellini, E., Lien, E. M., Pettitt, M. E., Callow, M. E., Callow, J. A., Conlan, S. L., Mutton, R. J., Clare, A. S., Kocijan, A., Donik, C., and Jenko, M. (2009) Fluorinated/siloxane copolymer blends for fouling release: chemi-

cal characterization and biological evaluation with algae and barnacles. *Biofouling*, **25**(6), 481-493.

77. Martinelli, E., Agostini, S., Galli, G., Chiellini, E., Glisenti, A., Pettitt, M. E., Callow, M. E., Callow, J. A., Graf, K., and Bartels, F. W. (2008) Nanostructured films of amphiphilic fluorinated block copolymers for fouling release application. *Langmuir*, **24**(22), 13138-13147.

78. Martinelli, E., Galli, G., Krishnan, S., Paik, M. Y., Ober, C. K., and Fischer, D. A. (2011) New poly(dimethylsiloxane)/poly(perfluorooctylethyl acrylate) block copolymers: structure and order across multiple length scales in thin films. *Journal of Materials Chemistry*, **21**(39), 15357-15368.

79. Martinelli, E., Hill, S. D., Finlay, J. A., Callow, M. E., Callow, J. A., Glisenti, A., and Galli, G. (2016) Amphiphilic modified-styrene copolymer films: Antifouling/fouling release properties against the green alga Ulva linza. *Progress in Organic Coatings*, **90**, 235-242.

80. Martinelli, E., Sarvothaman, M. K., Alderighi, M., Galli, G., Mielczarski, E., and Mielczarski, J. A. (2012) PDMS network blends of amphiphilic acrylic copolymers with poly(ethylene glycol)-fluoroalkyl side chains for fouling-release coatings. I. Chemistry and stability of the film surface. *Journal of Polymer Science, Part A: Polymer Chemistry*, **50**(13), 2677-2686.

81. Hawkins, M. L., Fay, F., Rehel, K., Linossier, I., and Grunlan, M. A. (2014) Bacteria and diatom resistance of silicones modified with PEO-silane amphiphiles. *Biofouling*, **30**(2), 247-258.

82. Hawkins, M. L., and Grunlan, M. A. (2012) The protein resistance of silicones prepared with a PEO-silane amphiphile. *Journal of Materials Chemistry*, **22**(37), 19540-19546.

83. Hawkins, M. L., Rufin, M. A., Raymond, J. E., and Grunlan, M. A. (2014) Direct observation of the nanocomplex surface reorganization of antifouling silicones containing a highly mobile PEO-silane amphiphile. *Journal of Materials Chemistry B*, **2**(34), 5689-5697.

84. Yasani, B. R., Martinelli, E., Galli, G., Glisenti, A., Mieszkin, S., Callow, M. E., and Callow, J. A. (2014) A comparison between different fouling-release elastomer coatings containing surface-active polymers. *Biofouling*, **30**(4), 387-399.

85. Bodkhe, R. B., Stafslien, S. J., Cilz, N., Daniels, J., Thompson, S. E. M., Callow, M. E., Callow, J. A., and Webster, D. C. (2012) Polyurethanes with amphiphilic surfaces made using telechelic functional PDMS having orthogonal acid functional groups. *Progress in Organic Coatings*, **75**(1-2), 38-48.

86. Bodkhe, R. B., Stafslien, S. J., Daniels, J., Cilz, N., Muelhberg, A. J., Thompson, S. E. M., Callow, M. E., Callow, J. A., and Webster, D. C. (2015) Zwitterionic siloxane-polyurethane fouling-release coatings. *Progress in Organic Coatings*, **78**, 369-380.

87. Webster, D. C., and Bodkhe, R. (2015) Functionalized Silicones with

Polyalkylene Oxide Side Chains, patent US9169359.

88. Galhenage, T. P., Webster, D. C., Moreira, A. M. S., Burgett, R. J., Stafslien, S. J., Vanderwal, L., Finlay, J. A., Franco, S. C., and Clare, A. S. (2016) Poly (ethylene) glycol-modified, amphiphilic, siloxane–polyurethane coatings and their performance as fouling-release surfaces. *Journal of Coatings Technology and Research*, **2**(14), 307-322.

89. Upadhyay, V., Galhenage, T., Battocchi, D., and Webster, D. (2017) Amphiphilic icephobic coatings. *Progress in Organic Coatings*, **112**, 191-199.

2

Polyetherimide Coatings for Corrosion Protection of Mild Carbon Steel: Effect of Residual Solvent and Temperature on Coating Performance

Tehsin Akhtar, Gisha Elizabeth Luckachan and Vikas Mittal*,**

Department of Chemical Engineering, The Petroleum Institute (part of Khalifa University of Science and Technology), Abu Dhabi, UAE

*Corresponding author: vik.mittal@gmail.com
**Current address: Bletchington, Wellington County, Australia

2.1 Introduction

High performance polymers are widely used as resistive coatings for the corrosion protection of metallic structures in aggressive environments (especially in marine conditions). The main role of these polymeric coatings is to protect the metal from external corrosive agents (oxygen, water, etc.) by acting as an effective barrier [1,2]. Besides that, polymers combine properties like thermal, mechanical and chemical stability, depending on their structure and morphology, which yield durable coatings with longer service life time of the coated metal substrates [1,3]. The effectiveness of these barrier coatings depends on the characteristics of the polymers to form defect-free layers by strong adhesion with metal surfaces.

Among different polymers used as protective coatings, polyetherimide (PEI) has drawn special attention because of its hydrophobicity, thermal stability and mechanical performance [4-8]. PEI is a high performance amorphous thermoplastic polymer with ether and imide groups. The presence of polar aromatic imide rings in the polymer structure enhances its adhesion to the metal surface. It is reported in the literature that the polymers with polar groups can behave as a base and oxides on the metal surface can behave as an acid in the Lewis sense, and this acid-base interaction provides good adhesion between the polymer and metal [9]. Besides that, PEI has a high glass transition temperature (T_g) of 216.7 °C and a low dielectric

Marine Coatings and Membranes, edited by Vikas Mittal
© 2019 Central West Publishing, Australia

constant [10-12] that provide a thermally stable coating at environmental conditions. Another interesting characteristic of PEI is its film forming property and flexibility to conform to the metal surface [4, 13], therefore, with the precise solvent choice and heating conditions, it can form defect-free, transparent and corrosion resistant coatings.

Choice of solvent or combination of solvents is important in producing high quality defect-free polymer coatings. Solvents can affect porosity, adhesion, appearance and gloss of the coatings [1,2,14,15]. Solvent retention in the coatings causes plasticization of the polymer, reduction in the T_g of the coatings as well as enhancement of influx of water and oxygen. The evaporation process of solvent also plays a role in determining the properties of the polymer coating and subsequently affects the morphology of the coatings to a large extent. Slow evaporating solvents allow sufficient time for wetting the substrate surface, thus, achieving good contact with the metal substrate for producing films with lesser defects and greater adhesion, along with preventing internal stresses which cause coating delamination [16,17]. However, high boiling point solvents can be difficult to remove from the coatings if the heating is not performed at suitable temperatures (preferably at temperatures higher than T_g of polymer used), which reduces the adhesion and barrier properties of the coatings, along with absorption of higher extent of moisture [1]. PEI is soluble in chlorinated solvents like dichloromethane (DCM) at room temperature and high boiling solvents like N-methyl pyrolidinone (NMP) and N,N-dimethylacetamide (DMAc) at elevated temperatures. Conceicao *et al.* [6,18] studied the influence of different coating processes, dip coating and spin coating, for preparing protective PEI coatings on magnesium AZ31 alloy using NMP and DMAc solvents [6,18]. The authors reported that the influence of residual solvent on the coating performance was tolerable for DMAc based coatings, however, it was detrimental for NMP coatings. Similar conditions were used by Zomorodian *et al.* [19] and Scharnagl *et al.* [14] for the fabrication of PEI coatings, and the values of coating impedances declined with time, reaching as low as $10^{4.5} - 10^6 \, \Omega \, cm^2$. Thus, for a barrier coating, the removal of residual solvent is essential to obtain good adhesion with the metal surface, thereby producing thin defect-free coatings suitable for aggressive environments. To analyze this quantitatively, a systematic analysis of the heating cycles to completely remove the residual solvent has been conducted in the current study, which enabled the fabrication of thin PEI coatings of <10 μm thickness, with enhanced barrier properties coupled with strong adhesion to the steel substrate. Interaction of PEI polymer with the metal substrate and solvent was analyzed by IR spectroscopy. The influence of different extents of residual solvent on the coating performance was studied by electrochemical impedance spectroscopy (EIS) and potentio-dynamic measurements.

2.2 Experimental

2.2.1 Materials

Carbon steel sheets of grade RST37-2 DIN 17100-80 were purchased from Qatar Steel Industries Factory, Qatar. Its composition in weight % was C (0.125), Mn (0.519), Si (0.016), P (0.014), S (0.005), Al (0.034) and Fe (99.287). Polyetherimide with melt flow index, 9 g/10 min (337 °C/6.6 kg) was purchased from Sigma-Aldrich. The solvents N,N-dimethylacetamide, N-methylpyrrolidone, dichloromethane and hydrochloric acid (37%) were obtained from Merck and used as received.

2.2.2 Substrate Preparation

Carbon steel coupons of dimensions 5 cm x 2 cm x 1.8 cm were pickled in concentrated hydrochloric acid until the mill scale was dissolved. These coupons were further ground and polished using three grades of sand paper (60, 150 and 180 grits) to remove the top layer of pickling compounds, oxides and hydroxides, thus, resulting in a smooth surface. The polished coupons were rinsed with distilled water and acetone, followed by drying in the oven at 80 °C before use.

2.2.3 Preparation of PEI Solutions

PEI pellets were dissolved in DCM at ambient temperature and in NMP and DMAc at 70 °C under magnetic stirring to produce homogeneous PEI solutions. PEI-DCM solutions were prepared with the following concentrations of PEI: 3, 5, 10, 15 and 20 wt%. 15 wt% solutions of PEI were also prepared in DMAc and NMP separately.

2.2.4 Preparation of PEI Coatings and Films

The cleaned steel coupons were dip-coated in the PEI solutions, using Qualtech Precision Dip Coater (model QPI 128). The immersion and withdrawal speeds were adjusted to 50 mm/min and 150 mm/min respectively, and the wetting time was fixed to 2 min to allow sufficient time for polymer to flow into the substrate surface irregularities.

Coatings with different solvents, PEI concentrations and heating conditions were formulated. PEI-DCM coatings with 3, 5, 10, 15 and 20 wt% PEI were fabricated at ambient temperature conditions. 15 wt% PEI-DCM, 15 wt% PEI-DMAc and 15 wt% PEI-NMP coatings were obtained by curing at 270 °C by following the sequential heating conditions: starting at 80 °C for 30 min, then at 120 °C, 150 °C, 180 °C, 210 °C and 240 °C for 15 min at each temperature, followed by final curing at 270 °C for 30 min. A heating sequence of 30 min at 80 °C, 15 min at each intermediate temperature and finally 30 min at the final temperature was also used to generate 15 wt% PEI-DMAc coatings at 150 °C and 240 °C. The coating thickness was measured using PosiTector 6000 coating thickness gauge.

Free standing PEI films for FTIR analysis were prepared by casting the 15 wt% PEI solutions in DCM, DMAC and NMP solvents on Teflon petri dishes, followed by curing at 270 °C using the same sequential heating conditions as the coatings.

For differential scanning calorimetry (DSC) analysis, 15 wt% PEI-DMAc films were prepared at different temperatures (80 °C, 120 °C, 150 °C, 180 °C, 210 °C, 240 °C and 270 °C) using a sequential heating of 30 min at 80 °C, 15 min at each intermediate temperature and 30 min at the final temperature. PEI-NMP films cured at 270 °C using the same sequential heating conditions were also prepared.

Finally, 15 wt% PEI-DMAc and 15 wt% PEI-NMP films and coatings were also prepared by 24 h drying at 80 °C.

2.2.5 Immersion Tests

Edges of the coated coupons were sealed using Nippon epoxy primer and dried for 24 h at room temperature. The coupons were immersed in 3.5 wt% NaCl solution, and the corrosion was monitored by the appearance of corrosion products or rust with time at room temperature.

2.2.6 Electrochemical Analysis

A three electrode cell (250 mL volume) with a platinum gauze counter electrode and a saturated calomel reference electrode (SCE) with bridge tube was used to perform electrochemical tests on the flat coated coupons. The coated coupon was the working electrode, and an area of 1 cm^2 was exposed to 3.5 wt% NaCl solution. All tests were carried out at room temperature by connecting the corrosion cell to BioLogic VMP-300 multipotentiostat (controlled by a computer running EC-Lab 10.40 software). Ultra-low current cables connected to the potentiostat were used for the accurate measurement of the current. This option included current ranges from 100 nA down to 100 pA, with additional gains extending the current ranges to 10 pA and 1 pA. The resolution on the lowest range was 76 aA. The open circuit potential (OCP) was measured for 5 min in order to allow the potential to stabilize before the electrochemical impedance and potentiodynamic polarization test. The impedance measurements were performed at amplitude of 20 mV over frequency range from 10^4 Hz to 10^{-2} Hz. The same software was used to simulate the impedance behavior of the samples. Polarization measurements were conducted by polarizing the working electrode from an initial potential of -250 mV up to a final potential of +250 mV as a function of open circuit potential, using a scan rate of 1.66 mV/s. The software was used to perform the fitting and evaluation of the corrosion rates using Faraday's equation for metal loss corrosion [2]. To study the reproducibility of the measurements, each set of experiments was repeated three times on newly coated samples.

2.2.7 Characterization Techniques

The glass transition temperature of pure PEI pellets and PEI films was determined using differential scanning calorimetry (DSC) employing Discovery DSC from TA instruments. The samples were placed in aluminum pans, and two heating (25 to 250 °C) and cooling cycles (250 to 25 °C) were performed at the rate of 10 °C/min under nitrogen flow, using a sample weight of 3-10 mg. The Fourier transform infrared spectroscopy of the coatings and films was performed using a Bruker VERTEX 70 FTIR spectrometer attached with a DRIFT accessory. IR acquisition was achieved by collecting 120 scans at a resolution of 4 cm^{-1} in the frequency range of 370 cm^{-1} to 4000 cm^{-1} (using OPUS software). Coated substrates were inspected closely for any signs of corrosion and local delamination with the help of digital microscope KH-7700 (Hirox Co. Ltd, USA). The coatings were scribed after completion of the immersion test and analyzed under optical microscope to observe any under-coating corrosion.

2.3 Results and Discussion

PEI coatings were prepared on carbon steel substrate by dip coating, using DCM, DMAc and NMP as solvents. The coatings prepared with 3, 5, 10, 15 and 20 wt% PEI in DCM and dried at room temperature were dense and transparent with non-porous morphology. During immersion in 3.5 wt% NaCl solution, severe pitting was observed on 3 and 5 wt% PEI coatings within 2 days. 10 wt% PEI coatings exhibited satisfactory performance for over 10 days, whereas the 15 wt% coatings had better protection even after 2 weeks of testing. The 20 wt% coatings disbonded and failed to provide protection because of contractive stresses due to excessive thickness of the coatings as well as expansive stresses when immersed in NaCl solution [16]. Thus, 15 wt% PEI coating was chosen for further analyses. The performance of 15 wt% PEI coating was, however, still not satisfactory, as adhesion loss and pitting corrosion were observed after 2 weeks of immersion in NaCl solution. The coatings did not have strong dry or wet adhesion. Therefore, 15 wt% PEI-DCM coatings were subjected to a high temperature curing (270 °C) using the sequential heating conditions (mentioned in the Experimental section) to reduce the susceptibility of the coatings to plasticization and to obtain good wet adhesion. These coatings were observed to be stable against corrosion for over 2.5 months, though they did not have the desired smooth finish due to blisters created by rapid solvent evaporation. Therefore, slow drying high boiling point solvents DMAc and NMP were analyzed for developing the stable PEI coatings for long time protection.

The morphology of PEI (15 wt%) coatings with DMAc and NMP was greatly affected by temperature and air humidity. Drying at 80 °C produced white, porous, non-uniform and non-adherent PEI coatings in both DMAc and NMP solvent formulations. However, immediate heat treatment of the just coated substrates induced phase inversion, in which morphology was controlled by solvent evaporation, thus, resulting in dense, clear and non-porous coatings. The PEI-DMAc coatings so produced had a smooth glossy finish and prevented corrosion up to 3 months in 3.5 wt% NaCl solution, as can be seen in Figure 2.1a and b. However, 15 wt% PEI-DMAc coatings prepared at 150 °C and 240 °C did not exhibit sufficient wet adhesion in immersion tests. Better performance of coatings cured at 270 °C was attributed to the removal of a large amount of residual solvent from coatings, thus, reducing the plasticization and increasing the T_g or rigidity of the coatings, which made them less vulnerable to water permeation [20-24].

The PEI-NMP coatings cured at 270 °C possessed good dry adhesion, but blistered swiftly. In addition, pitting and accumulation of rust were seen under the films which delaminated during the immersion test in less than a week, as seen in Figure 2.1c and d. This might have been due to the strong interaction of NMP with PEI, due to which the residual solvent was difficult to be removed from the coatings, even after treatment at 270 °C [18]. The residual solvent caused polymer plasticization, blistering, high coating capacitance (C_c) and, hence, poor coating performance in immersion tests. PEI-NMP coatings did not withstand 3.5 wt% NaCl solution for more than 5 days.

(a)   (b)   (c)   (d)

Figure 2.1 PEI-DMAc coatings prepared at 270 °C (a) before and (b) after 3 months of immersion in 3.5 wt% NaCl solution. PEI-NMP coatings prepared at 270 °C (c) before and (d) after 5 d of immersion in 3.5 wt% NaCl solution.

To analyze the interaction of PEI with solvents and metal surface as well as the effect of excess solvent on polymer structure, FTIR spectra of PEI coatings prepared with different solvents using different heating conditions were studied. The FTIR of pure PEI pellets in Figure 2.2a showed several peaks corresponding to asymmetric stretching of imide carbonyl groups (C=O) at 1740 cm^{-1}; symmetric stretching of imide carbonyl groups at 1786 cm^{-1}; C=C stretching at 1625 cm^{-1} and 1605 cm^{-1}; aromatic ring stretching at 1510 cm^{-1}, 1486 cm^{-1} and 1454 cm^{-1}; C-N-C stretching signals at 1390 cm^{-1} and 1371 cm^{-1}; ether linkage Ar-O-Ar (aromatic ether) at 1292 cm^{-1}, 1249 cm^{-1}, 1220 cm^{-1}, 1181 cm^{-1}, 1117 cm^{-1} and 1081 cm^{-1}, etc. [7,11,24-26]. IR spectra of PEI-DMAc coating dried at 80 °C showed a new peak at 1645 cm^{-1} and two shoulder peaks at 1708 cm^{-1} and 1343 cm^{-1}. The former is attributed to the C=O of DMAc and latter two shoulders are attributed to the changes in the imide group at the carbonyl and C-N-C bonds respectively (Figure 2.2b). After curing at 270 °C, the peak at 1645 cm^{-1} vanished completely and the two shoulder peaks at 1708 cm^{-1} and 1343 cm^{-1} formed well-defined signals (Figure 2.2c). On the

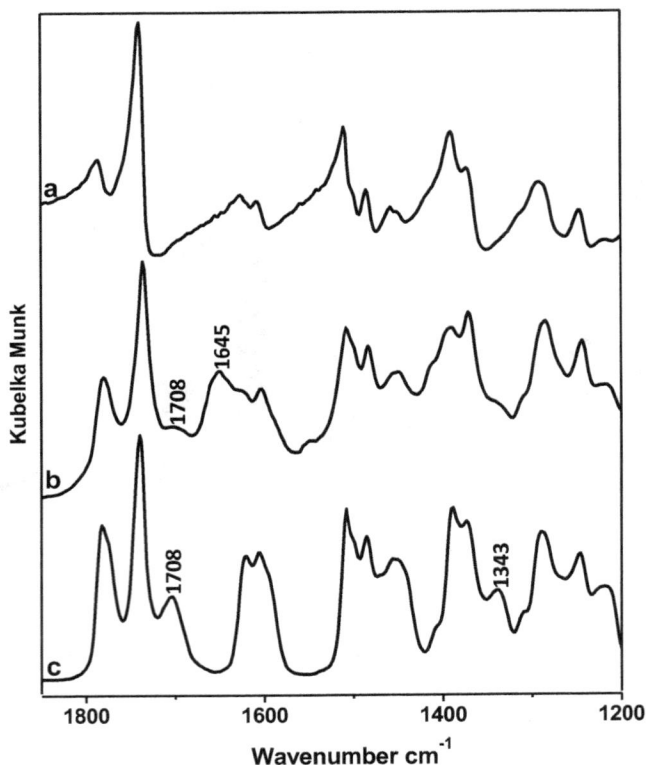

Figure 2.2 FTIR spectra of (a) pure PEI pellets, 15 wt% PEI-DMAc coatings prepared at (b) 80 °C and (c) 270 °C (c).

other hand, considering the PEI-NMP coatings, the contribution from solvent appeared as a sharp signal at 1683 cm^{-1} with a shoulder band attributed to imide carbonyl at 1713 cm^{-1} (Figure 2.3a). High temperature curing at 270 °C changed these bands to single well-separated broad bands centered at 1699 cm^{-1} and 1335 cm^{-1} (Figure 2.3b). Conceicao *et al.* [6] reported such bands in the IR spectra of spin coated PEI in NMP solvent after vacuum oven drying at 135 °C and suggested these peaks to result from the interaction of NMP with PEI polymer

[6]. However, in the present case, IR spectra of PEI coatings formulated with low boiling solvent DCM also showed these two well-distinguished bands at 1708 cm^{-1} and 1343 cm^{-1} after high temperature curing at 270 °C (Figure 2.3c). Therefore, to understand the nature of

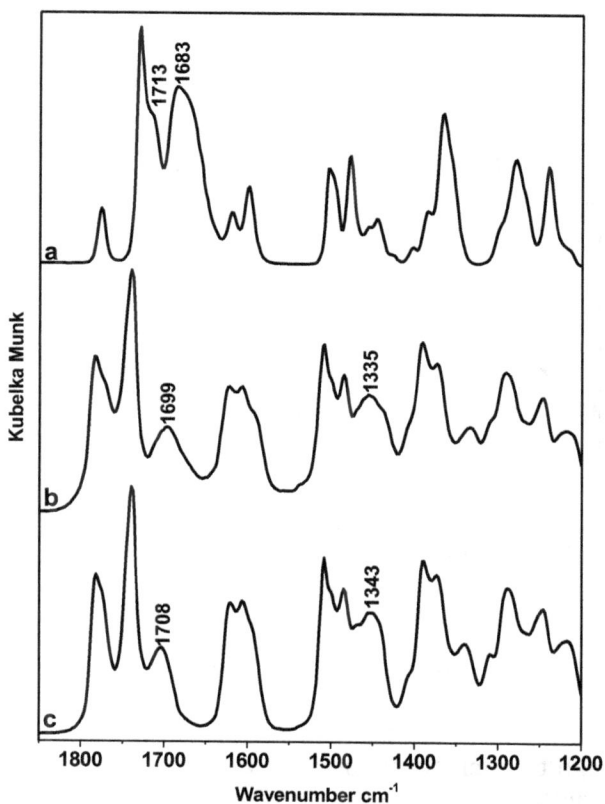

Figure 2.3 FTIR spectra of (a) PEI-NMP coating dried at 80 °C, (b) PEI-NMP coating cured at 270 °C and (c) PEI-DCM coating cured at 270 °C.

these bands, thin films of PEI in DMAc solvent were casted at 80 °C and 270 °C, and subjected to IR analysis, as shown in Figure 2.4. Like PEI-DMAc coatings, IR spectra of PEI-DMAc films dried at 80 °C showed solvent contribution at 1645 cm^{-1}. However, these did not show the bands attributed to the changes in the imide group at 1708 cm^{-1} and 1343 cm^{-1} even after curing at 270 °C (Figure 2.4a and b). Therefore, it can be confirmed that the new peaks at 1708 cm^{-1} and 1343 cm^{-1} appeared in the IR spectra of the PEI coatings due to the interaction of PEI with the steel substrate. It can be suggested that the functional groups of imide ring in polyetherimide, C=O and C-N-C groups, interacted with the steel surface extensively to form adhesive

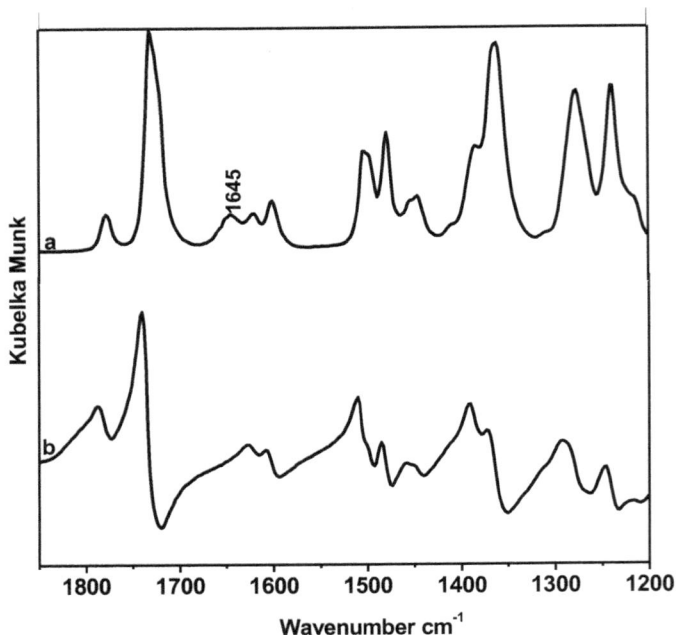

Figure 2.4 FTIR spectra of PEI-DMAc films prepared at (a) 80 °C and (b) 270 °C.

bonds at high temperatures (such as 270 °C). This observation was consistent with other studies reporting the bonding between polyimides and metals at room and elevated temperatures [27-31]. Atanasoska *et al.* [27] showed through XPS studies that the metals preferentially bind at polyimide carbonyl sites at room temperature. The authors reported that the metal became less selective at higher temperatures and attached to the imide ring functional group as well. The main site of reaction with the metal was the BPADA (bisphenol A dianhydride) portion of the polymer [27]. At room temperature, the metal transferred electrons to carbonyl carbon via oxygen in the highly conjugated π-system of the BPADA portion of polyimide, which was followed by localized interaction of the metal with carbonyl group leading to C=O bond cleavage at the surface and metal-O-C complex formation. The heat treatment of the metal-polyimide interface promoted an extended interaction of the metal surface with the polymer. Transfer of unshared electrons from nitrogen atoms of the polymer to the metal was proposed by Saarivirta *et al.* [30], in which the unshared electrons on the nitrogen atom interacted with the *d* orbitals of iron forming coordinate bonds [30]. Therefore, it can be confirmed that the carbonyl and imide group sites on the polyetherimide chains were involved in the reaction with mild steel substrate. At low temperatures, electron transfer from steel to the carbonyl group of the polymer resulted in the formation of Fe-O-C bonds. Fe-N bonds were produced by the interaction of electrons from nitrogen atoms with *d* orbitals of iron. Upon heat treatment, enhanced interaction of the polymer layer with iron substrate occurred. However, the interaction of polymer with solvent was obvious in PEI-NMP coatings that shifted the newly formed carbonyl and C-N-C bonds to lower regions of 1699 cm-1 and 1335 cm-1, as compared with 1708 cm-1 and 1343 cm-1 for PEI-DMAc and PEI-DCM coatings. The strong interaction of NMP with the polymer made it difficult to remove the solvent even after curing at 270 °C. It was believed that the interactions of steel with the polymer may also have been weakened due to the presence of NMP which might have interfered with adhesive bond formation. Weak adhesive bonds led to easy delamination and severe corrosion of PEI-NMP coatings during immersion tests (Figure 2.1d). The presence of residual solvent was further confirmed from the DSC and EIS results.

To develop coatings with high corrosion resistance, it is imperative to remove excess solvent molecules from coatings as high dielectric constant of solvents increases the coating capacitance, rendering them highly susceptible to further plasticization by increased water uptake. DMAc and NMP are solvents with high dielectric constants of 38.0 and 32.6 respectively [32,33]. Therefore, T_g of PEI films formed using DMAc and NMP solvents was measured to study the effect of solvent, in comparison with the pure PEI T_g of 216.7 °C. Films of 15 wt% PEI solutions in DMAc and NMP were prepared by heating at different conditions, as tabulated in Table 2.1. During first heating cycle of DSC analysis, weakly attached solvent molecules evaporated, however, a large extent of solvent remained in the polymer film, which reduced the T_g of PEI-DMAc films prepared at low temperatures. Complete removal of DMAc by curing at 270 °C increased T_g to 217.4 °C, which was very close to the value for pure PEI. The PEI-NMP films exhibited low T_g even when cured at 270 °C, indicating strong interaction of NMP with PEI, also observed earlier in the IR spectra.

These results confirmed DMAc to be a better solvent choice, as it could be effectively removed at high temperatures owing to its lower interaction with PEI.

Table 2.1 Glass transition temperature of PEI films with different solvents and heating conditions

Solvent	Film heating conditions		Glass transition temperature (T_g)	
	Tempera-ture** (°C)	Total heating time* (hours)	Heating curve 1	Heating curve 2
DMAc	80 °C	0.50	165.2 °C	212.3 °C
	120 °C	0.75	118.0 °C	195.8 °C
	150 °C	1.00	117.2 °C	196.9 °C
	180 °C	1.25	167.4 °C	180.6 °C
	210 °C	1.50	185.0 °C	199.1 °C
	240 °C	1.75	186.0 °C	211.7 °C
	270 °C	2.25	217.4 °C	216.8 °C
NMP	270 °C	2.25	-	174.8 °C

The findings from DSC were further confirmed from the coating capacitances obtained from EIS analysis shown in Figure 2.5. The impedance of 15 wt% PEI coatings, developed in DCM, DMAc and NMP solvents by heating at 80 °C and 270 °C, was measured in 3.5 wt% NaCl solution. The capacitance of the coatings was calculated by taking the inverse of impedance at the low frequency (0.16 Hz) region of the Bode plots [34], and the data is presented in Table 2.2. High temperature curing (270 °C) was observed to enhance the impedance of the coatings significantly (Figure 2.5a and b). The PEI-DCM and PEI-DMAc coatings had similar high impedance values of 4.46 x 10^{10} Ω cm^2 and 1.73 x 10^{10} Ω cm^2 respectively. The impedance value of NMP coatings was 1.38 x 10^9 Ω cm^2, an order of magnitude lower than that of PEI-DCM and PEI-DMAc coatings, indicating the strong interaction of NMP with PEI. Coating capacitance also exhibited the similar behavior. For PEI-DCM and PEI-DMAc coatings, the capacitance decreased to a similar value of the order of 10^{-10} F cm^2 indicating a complete removal of residual solvent. For PEI-NMP coating, the capacitance was still higher than the other two coatings showing the presence of residual solvent, even after the high temperature treatment, supporting the aforementioned IR and DSC results.

As significant corrosion protection of the substrate (over 3 months) was observed for 15 wt% PEI-DMAc coating cured at 270 °C

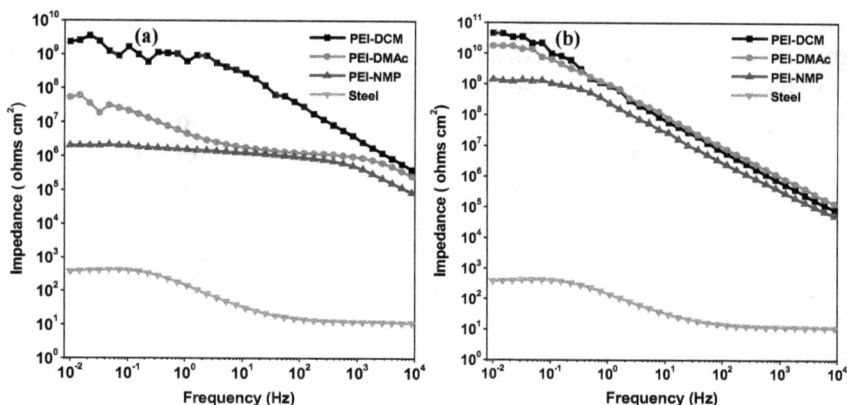

Figure 2.5 Bode plots of bare steel and PEI coatings prepared with different solvents at (a) 80 °C and (b) 270 °C.

Table 2.2 Impedance and capacitance values of different coatings calculated from Bode plots

Coatings		Impedance (ohm cm^2)	Capacitance (F cm^2)
PEI - DCM	80 °C	9.77 x 10^8	1.02 x 10^{-9}
	270 °C	4.07 x 10^9	2.45 x 10^{-10}
PEI - DMAc	80 °C	1.69 x 10^7	5.91 x 10^{-8}
	270 °C	4.46 x 10^9	2.24 x 10^{-10}
PEI - NMP	80 °C	1.81 x 10^6	5.52 x 10^{-7}
	270 °C	7.24 x 10^8	1.38 x 10^{-9}

as compared to PEI-NMP and PEI-DCM coatings during immersion test, a detailed study on the behavior of PEI-DMAc coating in 3.5 wt% NaCl solution was conducted by electrochemical experiments. EIS measurements of 15 wt% PEI-DMAc coatings of 7-8 μm thickness, recorded on a weekly basis, are presented in Figure 2.6. As can be observed from Bode plots, the PEI-DMAc coatings possessed a very high initial impedance modulus of the order of 10^{11} Ω cm^2 (Figure 2.6a). The high values of impedance at low frequencies indicate high degree of corrosion protection offered by the coatings [19,30]. Similar impedances are only described in the literature for much thicker and more complex coatings (e.g. 65 μm thick polyurethane and 200 μm epoxy-polyamide coatings on galvanized steel panels with initial low frequency impedance value of the order of 10^9-10^{10} Ω cm^2 [35,36]). The plateau in the Bode phase angle plot given in Figure 2.6b was nearly 90° over the complete range of measurement frequencies, which further confirmed the barrier effect or the high corrosion resistance of the PEI-DMAc coatings. The impedance of the coatings was independent of the immersion time and negligible changes in the electrochemical behavior were observed. The Bode plots recorded in the beginning of the test and at the end of 10 weeks were similar and no significant drop in the impedance values was seen during the entire observation period.

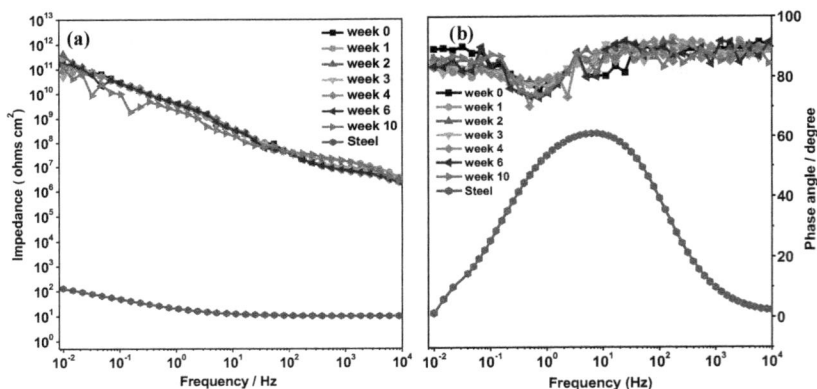

Figure 2.6 (a) Bode plots and (b) Bode phase angle plots of PEI-DMAc coatings at different times of immersion in 3.5 wt% NaCl solution.

The processes occurring in the coatings were simulated by an electrical network consisting of capacitors and resistors. The impedance plots were fitted to the equivalent circuit of Figure 2.7, and the values of different circuit components are shown in Table 2.3. R_s, R_c and R_{ct} represent solution resistance, coating/pore resistance and charge transfer resistance respectively. Constant phase elements, Q_c and Q_{dl}, are related to the coating capacitance and double layer capacitance (C_{dl}) respectively [37]. Z_W represents the Warburg impedance and accounts for mass transfer effects [38]. According to the previous studies, constant phase element permits the simulation of phenomena that deviate from a pure capacitive behavior [6,37,39]. Therefore, changes in the constant phase elements can be explained by evaluating the expression in terms of capacitance (C), as shown in Eq. 1 [40].

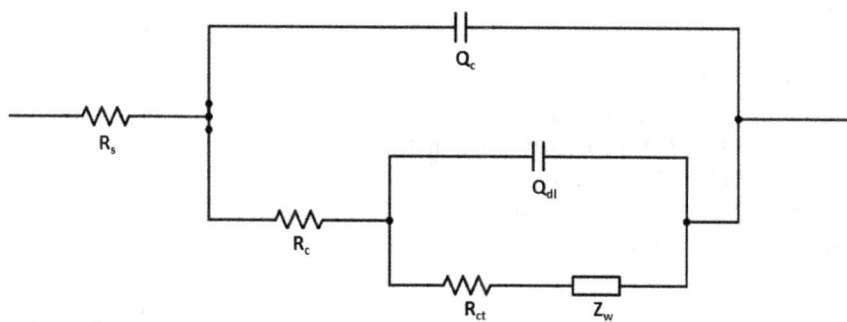

Rs - Solution resistance
Rc - Coating/pore resistance
Rct - Charge transfer resistance
Qc,Qdl - Constant phase elements
Zw - Warburg impedance

Figure 2.7 Circuit diagram used for fitting electrochemical data.

Table 2.3 Fitted parameters from Bode plots of PEI-DMAc coatings in comparison with bare steel

Week	Z (Ω cm^2)	Q_c		R_c (Ω cm^2)	Q_{dl}		R_{ct} (Ω cm^2)	W (Ω s$^{1/2}$)
		Y_{0c} (F s^{n-1})	n		Y_{0dl} (F s^{n-1})	n		
0	1.72×10^{11}	9.24×10^{-12}	1.000	9.12×10^{6}	3.88×10^{-11}	0.982	1.18×10^{10}	8.56×10^{10}
1	2.65×10^{11}	7.92×10^{-12}	1.000	2.61×10^{7}	4.19×10^{-11}	0.997	1.12×10^{10}	6.92×10^{10}
2	4.22×10^{11}	8.26×10^{-12}	1.000	1.27×10^{7}	3.98×10^{-11}	1.000	2.38×10^{10}	1.00×10^{11}
3	1.57×10^{11}	9.30×10^{-12}	1.000	1.25×10^{7}	3.96×10^{-11}	0.993	1.20×10^{10}	8.52×10^{10}
4	1.46×10^{11}	9.71×10^{-12}	1.000	7.77×10^{6}	3.54×10^{-11}	1.000	6.81×10^{10}	8.59×10^{10}
6	1.65×10^{11}	9.50×10^{-12}	0.960	7.57×10^{6}	3.88×10^{-11}	1.000	2.48×10^{10}	1.74×10^{10}
10	1.37×10^{11}	9.28×10^{-12}	0.978	7.06×10^{6}	4.27×10^{-11}	1.000	2.24×10^{10}	2.75×10^{8}
Steel	3.80×10^{2}	-	-	-	0.39×10^{-3}	0.818	1638	-

$$C = Y_0 \, (\omega_{max})^{\,n-1} \tag{Eq. 1}$$

where Y_0 is the magnitude of the constant phase element, ω_{max} is the frequency at which the imaginary impedance reaches a maximum for the respective time constant and 'n' is the exponential term of the CPE [40,41]. When 'n' is equal to 1, the CPE behaves as a pure capacitor, whereas for 'n' equal to 0, it represents a resistor. On the other hand, for 'n' equal to -1, it represents an inductor. Since constant 'n' in Table 2.3 is close to 1 over the complete measurement time, Q_c and Q_{dl} would be similar to pure capacitors, and their Y_0 constants follow the same trend as the capacitance in Eq. 2 [39,40].

$$C = \varepsilon\varepsilon_0(A/t) \tag{Eq. 2}$$

where ε is the dielectric constant of the material, ε_0 is the dielectric constant in vacuum, A is the area and t is the thickness of the coatings exposed to the corrosive medium [39]. Since capacitance is directly related to dielectric constant, the effect of residual solvent on the coating performance could be monitored from the Y_0 values [6,39]. A small drop in the Y_0 value of Q_c (Y_{0c}) from 9.24 x 10^{-12} F cm^2 to 7.92 x 10^{-12} F cm^2 and corresponding increase in the coating resistance (R_c) from 9.12 x 10^6 Ω cm^2 to 26.1 x 10^6 Ω cm^2 was observed in the first week of immersion (Table 2.3). This can be attributed to the removal of a small amount of residual solvent out of the coating in the initial hours of exposure to sodium chloride solution [18]. As the dielectric constant of the solvent is high, the movement of the solvent molecules out of the coating reduces the coating capacitance. Even though the amount of solvent in the coatings was very low due to high temperature curing, a small extent of entrapped solvent was nonetheless washed out upon exposure to the electrolyte, which reduced the plasticization of the film and diffusion coefficient of water further [18]. From second weak onwards, Y_{0c} values increased slightly, coupled with small decrease in pore resistance. This was due to the filling of the pores (which represent free volume, channels or discontinuities) of the coatings with electrolyte. This behavior is typical for most organic coatings [30,42]. No noticeable increase in the Y_{0c} values was observed towards the end of the measurement time, which indicated that the water uptake by the coatings was very slow (as the high T_g of the coatings restricted the movement of polymer chains and did not provide easy pathway for water influx [1]). The entered water can be visualized to be distributed throughout the coating, however, it did not reach the metal surface and did not result in corrosion. Though 10 weeks immersion decreased R_c slightly from 9.12 x 10^6 Ω cm^2 to 7.06 x 10^6 Ω cm^2, the resistance of the coating was still high to isolate the metal from aggressive environment. The Y_0 values of Q_{dl} (Y_{0dl}) were low as expected, and the charge transfer resistance (R_{ct}) had extremely high values of the order of 10^{10} Ω cm^2, showing the high degree of protection of the substrate. The Y_{0dl} and R_{ct} values stayed fairly constant throughout the test period, suggesting that no degradation of the coating occurred and no under-coating corrosion was generated. In addition, the coating impedance was much higher than any other PEI coating reported in literature [5,6,14,18,19,34]. Conceicao *et al.* [6,18] reported the maximum initial impedance for PEI coatings (of equal thickness) of the order of 10^9 Ω cm^2. However, in this study, the impedance of PEI coatings was of the order of $10^{11.5}$ Ω cm^2, two orders of magnitude higher than the impedance values reported by Conceicao *et al.* and four orders of magnitude higher than impedances obtained by Rout *et al.*, Zomorodian *et al.* and Scharnagl *et al.* [14,18,19,34]. Furthermore, the values of PEI coating impedances in these studies declined every week, reaching values as low as $10^{4.5}$ - 10^6 Ω cm^2, however, the PEI coatings prepared in this study were stable and had no impedance decline after long immersion time of 10 weeks. Such high and steady impedance values were attained due to the employment of the high temperature curing at 270 °C, which removed the residual solvent from the coatings and improved the metal/polymer interaction. In comparison, maximum temperature of 150 °C in combination with vacuum were used for drying of coatings in the literature studies, which may have resulted in partial removal of solvent, thus, producing coatings with higher capacitances and low impedances susceptible to corrosive attack and adhesion loss [5,6,14]. The coatings so produced blistered and corroded within a few days of immersion in sodium chloride solution. These results suggested that not only the high temperature treatment but also a systematic heating cycle from room to high temperature are essential for the fabrication of thin and stable PEI barrier coatings for achieving long term corrosion protection of metal substrates in contact with corrosive environments.

Tafel plots of polymer coatings were recorded by running the potentio-dynamic sweeps and are shown in Figure 2.8. The corrosion currents (I_{corr}), potentials (E_{corr}) and expected metal thickness loss in terms of corrosion rate (millimeters per year) are presented in Table 2.4. E_{corr} and cathodic as well as anodic currents were absent during

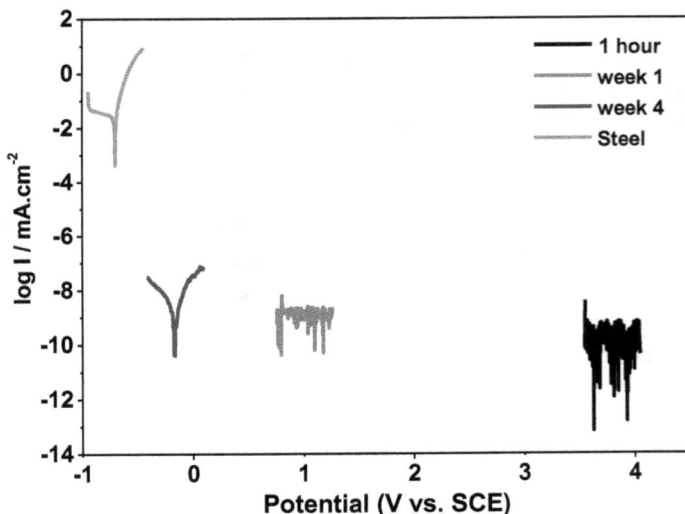

Figure 2.8 Tafel plots of PEI - DMAc coatings in comparison with bare steel.

Table 2.4 Fitted parameters from potentio-dynamic curves of PEI-DMAc coatings in comparison with bare steel

Time of immersion	Corrosion potential E_{corr} (mV vs. SCE)	Corrosion current I_{corr} ($\mu A.cm^{-2}$)	Corrosion rate (mmpy)
1 hour	-	2.78×10^{-7}	3.27×10^{-9}
1 week	-	3.00×10^{-6}	30.32×10^{-9}
4 weeks	-185.88	3.00×10^{-6}	35.32×10^{-9}
Steel	-702.32	44.477	0.523

the initial weeks of immersion, therefore, I_{corr} of the coatings during this time was calculated directly from the Tafel plots without performing any fitting analysis [34]. The PEI-DMAc coatings showed very low corrosion current density of 2.78×10^{-7} μA cm^{-2} after 1 hour immersion in 3.5 wt% NaCl solution indicating that the polymer was highly protective. Also, E_{corr} shifted towards negative values with time. After 4 weeks immersion, E_{corr} of PEI-DMAc coatings was recorded as -185.88 mV, which was still higher than the corrosion potential of bare steel (-702.32 mV). The I_{corr} value increased to 3.00×10^{-6} μA cm^{-2} during 1 week immersion and retained the same value after 4 weeks which was much lower than that of uncoated steel

(44.477 µA cm⁻²). Corrosion rate (CR) calculated from corrosion current, given in Table 2.4, also followed the same trend, maintaining a very low value of 35.32 x 10⁻⁹ mmpy after 4 weeks. These results indicated that PEI-DMAc coating (cured at 270 °C) provided excellent corrosion protection and persisted for long time in the corrosive environment. The shift of corrosion potential to negative values and very low corrosion current density suggested that the protection offered by PEI-DMAc coating was purely barrier in nature. The permeation of water through the PEI coatings was a very slow process, and as only minor corrosion currents were observed even after 4 weeks of immersion and stable high impedance was noted in Bode plots, the coatings proved their ability to provide protection to the underlying metal from corrosion.

The PEI-DMAc coatings were examined visually for any signs of corrosion after 10 weeks of immersion test in 3.5 wt% NaCl solution. No damage or delamination was observed and the metal surface was still covered with the insulating polymer, which prevented it from the corrosive attack owing to the barrier effect (Figure 2.9a). Similarly, pitting attack was also resisted by the coatings. Even under-coating corrosion was not detected on the steel surface, which was exposed by making X scribes, as shown in Figure 2.9b. No influx of water and loss of adhesion through the X scribed region were noted, even after 6 weeks of exposure to corrosive electrolyte, showing excellent adhesion and protection offered by the polymer coating (Figure 2.9c).

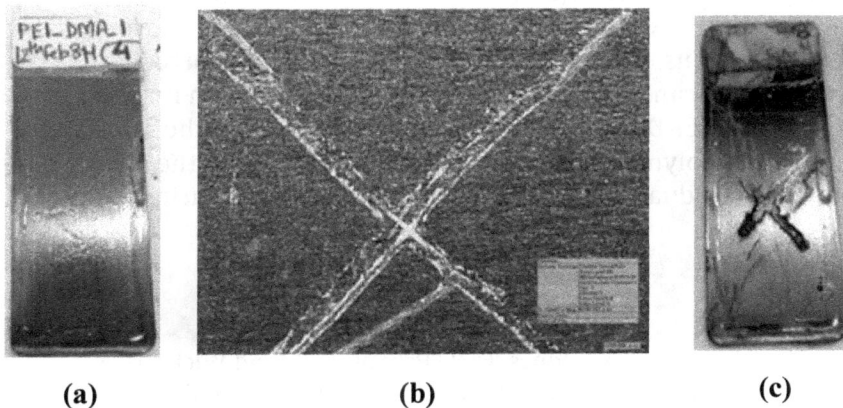

(a)   (b)   (c)

Figure 2.9 Appearance of PEI-DMAc coating after 10 weeks of immersion in 3.5 wt% NaCl solution: (a) digital image, (b) optical image of scribed area and (c) digital image of the scribed area after 6 weeks of immersion in 3.5 wt% NaCl solution.

2.4 Conclusions

Effect of residual solvent (DMAc, NMP and DCM) and temperature on the corrosion resistance of high performance polymers was studied by fabricating polyetherimide coatings on carbon steel substrate. High performance PEI coatings were obtained by following a heat treatment cycle involving gradual heating up to 270 °C. The 15%PEI-DMAc coatings resisted corrosion over 10 weeks in 3.5 wt% NaCl solution. It was attributed to the strong adhesive bond formation, observed in IR spectra, namely the Fe-O-C and Fe-N bonds, due to electron transfer from steel to the imide groups of the polymer. Although PEI-NMP coatings (cured at 270 °C) also formed strong adhesive bonds with metal, the residual solvent in the coating (confirmed from the low T_g values) and high coating capacitance plasticized the polymer and allowed easy penetration of corrosive agents through the coating, thus, resulting in delamination within 5 days of immersion. High degree of protection offered by PEI-DMAc coatings resulted in low frequency impedance of the order of 10^{11} Ω cm^2 in the Bode plot, with 7 to 8 µm coating thickness. During the entire observation period of 10 weeks, no significant drop in the low frequency impedance was observed, indicating the persistence of coating stability for long time. Moreover, the R_c and R_{ct} values were in the order of 10^6 Ω cm^2 and 10^{10} Ω cm^2 respectively, showing the coating efficiency in providing corrosion resistance. A very low corrosion rate of 35.32 x 10^{-9} mmpy was observed after 4 weeks immersion in sodium chloride solution. No influx of water or loss of adhesion occurred through the X scribed regions, even after 6 weeks of exposure to corrosive electrolyte. Thus, it can be concluded that the high corrosion resistance and stability of PEI-DMAc coatings were obtained due to the strong adhesion of the polymer with steel surface, supported by the effective removal of residual solvent during high temperature curing at 270 °C.

References

1. Wicks, Jr., Z. W., Jones, F. N., Pappas, S. P., and Wicks, D. A., (2007) *Organic Coatings: Science and Technology*, John Wiley & Sons, USA.
2. Ahmad, Z. (2006) *Principles of Corrosion Engineering and Corrosion Control*, Butterworth-Heinemann, UK.
3. Makhlouf, A. S. H. (2014) *Handbook of Smart Coatings for Materials Protection*, Elsevier, USA.
4. Dennis, R. V., Viyannalage, L. T., Gaikwad, A. V., Rout, T. K., and

Banerjee, S. (2013) Graphene nanocomposite coatings for protecting low-alloy steels from corrosion. *American Ceramic Society Bulletin*, **92**, 18-24.

5. Gaikwad, A. V., and Rout, T. K. (2011) In situ synthesis of silver nanoparticles in polyetherimide matrix and its application in coatings. *Journal of Materials Chemistry*, **21**, 1234-1239.

6. Conceicao, T. F., Scharnagl, N., Blawert, C., Dietzel, W., and Kainer, K. (2010) Corrosion protection of magnesium alloy AZ31 sheets by spin coating process with poly (ether imide) [PEI]. *Corrosion Science*, **52**, 2066-2079.

7. Chen, B.-K., Su, C.-T., Tseng, M.-C., and Tsay, S.-Y. (2006) Preparation of polyetherimide nanocomposites with improved thermal, mechanical and dielectric properties. *Polymer Bulletin*, **57**, 671-681.

8. Bijwe, J., Indumathi, J., Rajesh, J. J., and Fahim, M. (2001) Friction and wear behavior of polyetherimide composites in various wear modes. *Wear*, **249**, 715-726.

9. Mittal, K. L. (2012) *Adhesion Aspects of Polymeric Coatings*, Springer Science & Business Media, Germany.

10. Xian, G., and Zhang, Z. (2005) Sliding wear of polyetherimide matrix composites: I. Influence of short carbon fibre reinforcement, *Wear*, **258**, 776-782.

11. Amancio-Filho, S., Roeder, J., Nunes, S., Dos Santos, J., and Beckmann, F. (2008) Thermal degradation of polyetherimide joined by friction riveting (FricRiveting). Part I: Influence of rotation speed. *Polymer Degradation and Stability*, **93**, 1529-1538.

12. Liu, T., Tong, Y., and Zhang, W.-D. (2007) Preparation and characterization of carbon nanotube/polyetherimide nanocomposite films. *Composites Science and Technology*, **67**, 406-412.

13. Kostina, Y. V., Bondarenko, G., Alent'Ev, A. Y., and Yampol'Skii, Y. P. (2007) Effect of structure and conformational composition on the transport behavior of poly (ether imides). *Polymer Science Series A*, **49**, 77-88.

14. Scharnagl, N., Blawert, C., and Dietzel, W. (2009) Corrosion protection of magnesium alloy AZ31 by coating with poly (ether imides)(PEI). *Surface and Coatings Technology*, **203** 1423-1428.

15. Menut, P., Pochat-Bohatier, C., Deratani, A., Dupuy, C., and Guilbert, S. (2002) Structure formation of poly (ether-imide) films using nonsolvent vapor induced phase separation: relationship between mass transfer and relative humidity. *Desalination*, **145**, 11-16.

16. Negele, O., and Funke, W. (1996) Internal stress and wet adhesion of organic coatings. *Progress in Organic Coatings*, **28**, 285-289.

17. Perera, D. Y. (1996) On adhesion and stress in organic coatings, *Progress in organic coatings*, **28**, 21-23.

18. da Conceicao, T. F., Scharnagl, N., Dietzel, W., and Kainer, K. (2011) Corrosion protection of magnesium AZ31 alloy using poly (ether

imide)[PEI] coatings prepared by the dip coating method: Influence of solvent and substrate pre-treatment. *Corrosion Science*, 53338-346.

19. Zomorodian, A., Garcia, M., e Silva, T. M., Fernandes, J., Fernandes, M., and Montemor, M. (2013) Corrosion resistance of a composite polymeric coating applied on biodegradable AZ31 magnesium alloy. *Acta Biomaterialia*, **9**, 8660-8670.

20. Mittal, K. L. (1976) Adhesion measurement of thin films. *Active and Passive Electronic Components*, **3**, 21-42.

21. Wang, D., Li, K., and Teo, W. (1999) Phase separation in polyetherimide/solvent/nonsolvent systems and membrane formation. *Journal of Applied Polymer Science*, **71**, 1789-1796.

22. Itta, A. K., Tseng, H.-H., and Wey, M.-Y. (2010) Effect of dry/wet-phase inversion method on fabricating polyetherimide-derived CMS membrane for H_2/N_2 separation. *International Journal of Hydrogen Energy*, **35**, 1650-1658.

23. Ren, J., Zhou, J., and Deng, M. (2010) Morphology transition of asymmetric polyetherimide flat sheet membranes with different thickness by wet phase-inversion process. *Separation and Purification Technology*, **74**, 119-129.

24. Kurdi, J., and Kumar, A. (2005) Structuring and characterization of a novel highly microporous PEI/BMI semi-interpenetrating polymer network. *Polymer*, **46**, 6910-6922.

25. Ratta, V. (1999) Polyimides: Chemistry & structure-property relationships – Literature review. Faculty of Virginia Polytechnic Institute and State University, USA, pp. 3-28.

26. Chen, H.-L., You, J.-W., and Porter, R. S. (1996) Intermolecular interaction and conformation in poly (ether ether ketone)/poly (ether imide) blends - An infrared spectroscopic investigation. *Journal of Polymer Research*, **3**, 151-158.

27. Atanasoska, L., Anderson, S. G., Meyer, H., and Weaver, J. (1990) XPS study of chemical bonding at polyimide interfaces with metal and semiconductor overlayers. *Vacuum*, **40**, 85-90.

28. Freilich, S., and Ohuchi, F. (1987) Reactions at the polyimide-metal interface. *Polymer*, **28**, 1908-1914.

29. Ramos, M. M. (2002) Theoretical study of metal–polyimide interfacial properties. *Vacuum*, **64**, 255-260.

30. Huttunen-Saarivirta, E., Yudin, V., Myagkova, L., and Svetlichnyi, V. (2011) Corrosion protection of galvanized steel by polyimide coatings: EIS and SEM investigations. *Progress in Organic Coatings*, **72**, 269-278.

31. Warren, G., Sharma, R., Nikles, D., Hu, Y., and Street, S. (1999) Amine quinone polyurethane polymers for improved performance in advanced particulate media. *Journal of Magnetism and Magnetic Materials*, **193**, 276-278.

32. Thitiwongsawet, P., Sae-Lee, P., Kwanduen, P., Chinpa, W., and Supaphol, P. (2011) Electrospun poly (bisphenol A-co-4-nitrophthalic anhydride-co-1, 3-phenylenediamine) fibers: Preparation and potential for use in filtration applications. *Songklanakarin Journal of Science and Technology*, **33**, 315-323.

33. Hammerich, O., and Lund, H. (2000) *Organic Electrochemistry*, CRC Press, USA.

34. Rout, T. K., and Gaikwad, A. V. (2015) In-situ generation and application of nanocomposites on steel surface for anti-corrosion coating. *Progress in Organic Coatings*, **79**, 98-105.

35. Gonzalez-García, Y., González, S., and Souto, R. (2007) Electrochemical and structural properties of a polyurethane coating on steel substrates for corrosion protection. *Corrosion Science*, **49**, 3514-3526.

36. Gonzalez, S., Gil, M., Hernández, J., Fox, V., and Souto, R. (2001) Resistance to corrosion of galvanized steel covered with an epoxy-polyamide primer coating. *Progress in Organic Coatings*, **41**, 167-170.

37. Hirschorn, B., Orazem, M. E., Tribollet, B., Vivier, V., Frateur, I., and Musiani, M. (2010) Determination of effective capacitance and film thickness from constant-phase-element parameters. *Electrochimica Acta*, **55**, 6218-6227.

38. Park, J. H., and Park, J. M. (2014) Electrophoretic deposition of graphene oxide on mild carbon steel for anti-corrosion application. *Surface and Coatings Technology*, **254**, 167-174.

39. Moreno, C., Hernández, S., Santana, J., González-Guzmán, J., Souto, R., and Gonzalez, S. (2012) Characterization of water uptake by organic coatings used for the corrosion protection of steel as determined from capacitance measurements. *International Journal of Electrochemical Science*, **7**, 8444-8457.

40. Hsu, C. H., and Mansfeld, F. (2001) Technical note: Concerning the conversion of the constant phase element parameter Y0 into a capacitance. *Corrosion*, **57** 747-748.

41. Zheludkevich, M., Serra, R., Montemor, M., Yasakau, K., Salvado, I. M., and Ferreira, M. (2005) Nanostructured sol–gel coatings doped with cerium nitrate as pre-treatments for AA2024-T3: corrosion protection performance. *Electrochimica Acta*, **51** 208-217.

42. Singh, R. R., Banerjee, P. C., Lobo, D. E., Gullapalli, H., Sumandasa, M., Kumar, A., Choudhary, L., Tkacz, R., Ajayan, P. M., and Majumder, M. (2012) Protecting copper from electrochemical degradation by graphene coating. *Carbon*, **50** 4040-4045.

3

Anti-biofouling by Hydrophilic Polymer Brushes and Force Measurement of Cypris

Motoyasu Kobayashi,* Yuka Yamaguchi and Shouhei Shiomoto

School of Advanced Engineering, Kogakuin University, 2665-1 Nakano-cho Hachioji, Tokyo 192-0015, Japan

Corresponding author: motokoba@cc.kogakuin.ac.jp

3.1 Introduction

Any surface suffers from the settlement of marine organisms, such as bacteria, planktons, microorganisms and barnacles, which poses serious problems for the marine industry [1]. In particular, the sessile process and permanent adhesion mechanism of barnacles and their life cycles have attracted much attention because of the development of useful anti-biofouling materials and coatings [2-3]. An adult barnacle produces a planktonic nauplius which grows to a non-feeding cyprid larva within one or two weeks. The cypris larva begins exploring suitable locations for settlement by repeated tentative touches using the adhesive discs on its paired tentacles and, eventually, settles on a certain position on the substrate surface to metamorphose into the juvenile barnacle [4]. During the surface exploration stage, the cypris probes the surface morphology and physicochemical properties of the substrate in seawater by using its tentacles, similar to walking motion, leaving a "footprint protein" on the surface [5]. The footprint protein secreted from the attachment disk of the tentacles forms temporary anchoring points for the migrating cyprids and acts as settlement cues [6] because of the settlement-inducing protein complex (SIPC) [7]. Over the two decades, the anti-biofouling properties of hydrophilic polymer brushes have been reported in several research studies. In general, the polymer brushes are the surface-tethered polymers, which can modify the physical and chemical properties of the material surface owing to their functional groups [8]. In particular, the graft density of the brushes has been significantly

Marine Coatings and Membranes, edited by Vikas Mittal
© 2019 Central West Publishing, Australia

increased owing to "grafting-from" method supported by the controlled polymerization technique. When the densely-grafted polymer brush is immersed in a good solvent for the polymer, the large osmotic pressure results in a stretched chain structure [9], like hair and tooth brushes. For example, hydrophilic polyelectrolyte brushes in water form a swollen brush structure, which repel the oil droplet to promote the detachment of the oil contaminats or proteins from the surface, resulting in the antifouling and biofouling surfaces. This chapter describes in detail the synthesis of polymer brushes and their biofouling properties preventing the settlement of cypris larvae in seawater.

On the other hand, force measurement technique has also been advanced to observe the adhesive interaction between the substrate surface with various hydrophilicities and the proteins, cells and sessile organisms. Direct measurement of the temporary adhesion strength of live cypris larvae was first reported by Yule and coworkers using a sensitive electro-microbalance in 1983 [10,11]. Recently, scanning probe microscopy (SPM) technique, in particular, atomic force microscopy (AFM), has been used to measure the adhesion force of the footprint protein [12] or the tentacles of a live cypris [13] in seawater on the chemically-modified surface with well-designed hydrophobic or hydrophilic functional molecules. Therefore, the recent studies on the adhesion force measurement of the live cypris and footprint protein are also described in this chapter.

3.2 Preparation of Polymer Brushes

Preparation strategies of the polymer brushes are usually classified into "grafting-to" and "grafting-from" methods, as shown in Figure 3.1. The former involves the adsorption of block copolymer or end-functionalized polymer chains having functional groups that selectively interact with the substrate. The process is simple and applicable for the modification of a large surface area. However, the graft density of the resulting brush becomes low because the number of polymer chains per unit area (graft density) is limited by the size exclusion effect. On the other hand, the "grafting-from" method is able to form densely-grafted brushes as the monomers with smaller molecular size are polymerized from the surface. However, it is technically difficult to fabricate the brushes on the surface with a large area homogeneously. Although both "grafting-to" and "grafting-from" methods are extensively used for the fabrication of anti-biofouling surfaces, this chapter mainly describes the "grafting-from" method combined with controlled radical polymerization.

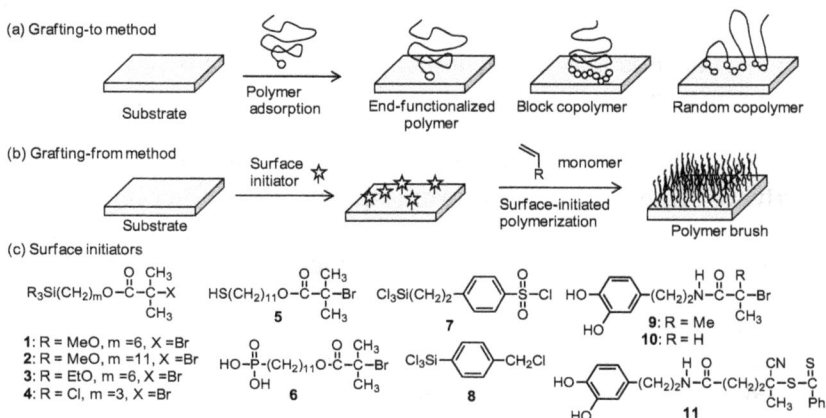

Figure 3.1 General scheme for the preparation of polymer brushes based on (a) "grafting-to" method and (b) "grafting -from" method, and (c) typical chemical structure of surface initiators used for the surface initiated controlled radical polymerization..

In general, the "grafting-from" method consists of two steps; the immobilization of the surface initiator on the substrate and surface-initiated polymerization. Surface initiator molecules consist of three parts: an anchoring group, a spacer and an initiation functional group, as illustrated in Figure 3.1(c). The surface initiator plays an important role as the anchoring point of the grafting polymer chains. As an anchoring group, the appropriate functional group is used to form chemical interaction with the substrate material. Alkoxy silanes (**1~4**) and silyl halides (**7,8**) are useful for binding with glass, quartz, silicon, silicate and indium tin oxide through Si-O bonds. Phosphoric acid (**6**) and catechol (**9,10**) can coordinate strongly with metal oxides such as stainless steel, iron oxide, alumina, zirconia and titanium oxide. Thiol (SH) groups (**5**) are used for gold surfaces. Therefore, the anchoring functional group is determined by the substrate. Instead of monolayer of surface initiators, polydopamine and barnacle cement thin films were also used for the initiator anchoring layer for the surface-initiated atom transfer radical polymerization (SI-ATRP) [14].

The other end of the molecule anchored on the substrate surface can initiate the polymerization of vinyl monomers. Various types of initiator head groups have been proposed in response to chain growth processes. Recently, most of the anti-biofouling polymer brushes have been prepared by SI-ATRP, therefore, α-bromo or α-chloro compounds having a carbon-halogen (C-X) bond are often used. Reversible addition-fragmentation chain transfer (RAFT) polymerization is another promising controlled radical polymerization method and has been often used for the preparation of polymer brushes initiated with **11**. The basic principle of ATRP is a reversible activation-dormancy process. The dormant species with the C-X bond is activated to produce carbon radicals when the homolytic scission of the C-X bond takes place tentatively by the oxidation of a copper halide CuX catalyst. The active radical species is capable of reacting with the monomer for propagation, but is immediately deactivated by the reproduction of the C-X bond. As the equilibrium between the dormant and active species is shifted to the dormant side, undesirable radical coupling reactions are restricted to provide quantitative initiator efficiency, sustainable chain growth without significant differences in the propagation rate of each chain and a polymer with narrow molecular weight dispersity. Thus, the outcome is completely different point from the conventional free radical polymerization. It is difficult to obtain high-density brushes by conventional free radical polymerization due to the low efficiency of the initiator, frequent termination by radical coupling and undesired chain transfer reactions.

3.3 Anti-biofouling Properties of Polymer Brushes

In general, marine fouling begins with the formation of an initial conditioning film [15] by the absorption of protein, polysaccharides and organo-compounds. It proceeds to the primary colonization stage caused by the adsorption and growth of bacteria as well as bio-film formation, which induces the secondary colonization by microorganism settlement and slime formation. Consequently, macro-organism settlement and growth eventually take place to constitute the tertiary colonization stage. To achieve the anti-biofouling surface modification, various types of functional polymer brushes have been proposed based on the strategies such as "fouling resistant", "fouling release" and "fouling degradation" [2]. The inspiration for fundamental fouling resistant strategy has stemmed from the use of hydrophilic poly(ethylene glycol) (PEG) in biomedical applications, which resists protein adsorption and cell adhesion. Therefore, the surface coating and grafting approaches of hydrophilic polymers are expected to result in the inhibition of marine biofouling. Fluoropolymers and silicones are used as fouling release polymers, as their low surface energy reduces the adhesion strength between fouling organisms and the material surface to promote easy detachment and cleaning [16-17]. Cationic polymers with quaternary ammonium compounds have been widely employed as antibacterial and fouling degrading polymers as they disrupt cellular membranes of biofoulant, leading to death [18-19]. Overall, a significant number of anti-biofouling polymer brushes with various chemical structures have been developed so far, as described in different literature reviews [2-3]. This chapter

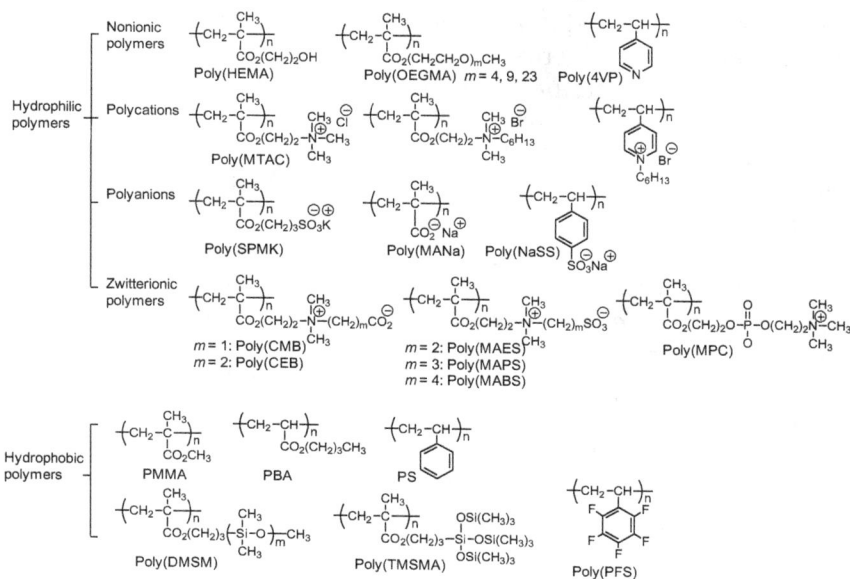

Figure 3.2 Chemical structure and abbreviations of polymer brushes used for assay of marine organism fouling. HEMA = 2-hydroxyethyl methacrylate, MATC = 2-(methacryloyloxy)ethyltrimethylammonium chloride, SPMK = 3-sulfopropyl methacrylate potassium salt, MANa = methacrylic acid sodium salt, PSS = poly(4-syrenesulfonic acid sodium salt), CMB = N-(2-methacryloyloxyethyl)-N,N-dimethyl-N-methylcarboxybetaine, CEB = N-(2-methacryloyloxyethyl)-N,N-dimethyl-N-ethylcarboxybetaine, MAES = 3-[dimethyl(2'-methacryloyloxyethyl)ammonio]ethanesulfonate, MAPS = 3-[dimethyl(2'-methacryloyloxyethyl)ammonio]propanesulfonate, MABS = 3-[dimethyl(2'-methacryloyloxyethyl)ammonio]butanesulfonate, MPC = 2-methacryloyloxyethyl phosphorylcholine, PMMA = poly(methyl methacrylate), PS = polystyrene, PBA = poly(butyl acrylate), DMSM = 3-[poly(dimethylsiloxy)silyl]propyl methacrylate, TMSM = 3-[tris(trimethylsiloxy)silyl]propyl methacrylate and PFS = 2,3,4,5,6-pentafluorostyrene.

highlights the functional polymer brushes (Figure 3.2) used in settlement experiments of cypris larvae.

Poly[2-(methacryloyloxy)ethyl phosphorylcholine] (Poly(MPC)) has been well known as an excellent biocompatible [20] polyzwitterion with antithrombogenicity [21] in the biomaterial research community, since its synthesis for the first time in 1990 by Ishihara *et al.* [22]. In general, surface grafting of ion-containing polymers affords a hydrophilic surface. In particular, the poly(MPC) brushes provide a superhydrophilic surface with an extremely low water contact angle below 5°, which prevents the adsorption of protein, cells [23] and oily contamination [24-25] in aqueous media. A similar hydrophilic surface can be obtained by the surface grafting of other polyzwitterions, such as poly(sulfobetaine) [26] and poly(carboxybetaine) consisting of *N*-(2-methacryloyloxyethyl)-*N,N*-dimethyl-*N*-methylcarboxybetaine (CMB) [27]. Aldred *et al.* [28] prepared polyzwitterion brushes by the SI-ATRP of 3-[dimethyl(2'-methacryloyloxyethyl)ammonio]propanesulfonate (MAPS) and *N*-(2-methacryloyloxyethyl)-*N,N*-dimethyl-*N*-ethylcarboxybetaine (CEB) on a glass substrate and observed the tracking behavior of the cypris larva of the barnacle *Amphibalanus amphitrite* (*Balanus amphitrite*). The brush surfaces exhibited high resistance against the settlement of cyprids. Interestingly, different exploring behaviors of cyprids were observed. The cyprids explored the poly(MAPS) brush surface attempting surface attachment but did not settle, whereas the cyprids did not attempt to explore the poly(CEB) brush and left the surface quickly.

Yang *et al.* [29] prepared ionic and non-ionic polymer brushes by the SI-ATRP of MAPS, 2-(methacryloyloxy)ethyltrimethylammonium chloride (MTAC), 4-syrenesulfonic acid sodium salt (NaSS), 2-hydroxyethyl methacrylate (HEMA) and 2,3,4,5,6-pentafluorostyrene (PFS) on glass substrate to investigate the antifouling effect of polymer brushes against marine organisms, by performing settlement tests with barnacle cypris larvae *Amphibalanus amphitrite* as well as bovine serum albumin (BSA), gram-positive *Staphylococcus aureus*, and gram-negative bacterium *Pseudomonas sp.* NCIMN 2021 [29]. The authors reported that the non-ionic poly(HEMA), zwitterionic poly(MAPS) and cationic poly(MTAC) brushes exhibited better antifouling performance against the adhesion of bacteria and settlement of cypris larvae. In contrast, the anionic poly(NaSS) brush surface could not deter the adsorption of bacteria and cypris settlement. The poly(PFS) brush showed effective reduction in protein absorption and bacterial adhesion due to the fouling-release effect based on the low surface energy, however, settlement of cypris larvae was observed. It was concluded that the ability to inhibit fouling by bacteria and cyprids was in the order of poly(HEMA) ~ poly(MAPS) > poly(MTAC) >> poly(PFP) > poly(NaSS).

Higaki *et al.* [30] also investigated the antifouling potential of polymer brushes against marine organisms by performing settlement tests with barnacle cypris larvae *Amphibalanus amphitrite* and blue mussel larvae in their adhesion period as well as a marine bacteria colonization test. Hydrophilic and hydrophobic polymer brushes were prepared on silicon wafers by the SI-ATRP of MTAC, MPC, MAPS, N-(2-methacryloyloxyethyl)-*N,N*-dimethyl-*N*-ethylcarboxybetaine (MAES), 3-[dimethyl(2'-methacryloyloxyethyl)ammonio]propane-sulfonate (MABS), poly(methyl methacrylate) (PMMA), polystyrene (PS), poly(*n*-butyl acrylate) (PBA), 3-[tris(trimethylsiloxy)silyl]propyl methacrylate (TMSM) and 3-[poly(dimethylsiloxy)silyl]propyl methacrylate (DMSM). On hydrophobic PBA, poly(TMSM) and poly(DMSM) brush surfaces, the number of settled cypris larvae increased with time; in particular, a significant increase was observed after 7 days. Although polysiloxane elastomers are known as fouling-release polymers due to their low surface energy and low modulus, the fouling-release effect was not observed for the siloxane-containing poly(DMSM) brush. On the other hand, zwitterionic poly(MPC), poly(MAES), poly(MAPS) and poly(MABS) brushes exhibited excellent prevention of cyprids settlement during the 10 days assay. Similar effective antifouling properties of polyzwitterion brushes were observed for mussel larvae and bacteria, however, the cationic poly(MTAC) brushes allowed mussel larvae settlement and bacteria adhesion, same as hydrophobic PS and PMMA brushes. This was an unexpected behavior as the cationic polymers bearing the quaternary ammonium group are among the best known antimicrobial polymers [31,32]. Actually, antibacterial effects were observed on the positively charged polymer surfaces prepared by quaternary poly(*N,N*-dimethylaminoethyl methacrylate) [33] and poly(4-vinyl pyridine) [34] brushes. Therefore, bio-films might have formed on the poly(MTAC) brush surface by the deposition of dead bacteria, thus, inducing biofouling.

The presented literature makes it possible to conclude that zwitterionic polyelectrolyte brushes exhibit an excellent antifouling character inhibiting the settlement of cypris larvae. Superhydrophilicity is considered as one of the key factors to prevent settlement. Additional possible factors are the swollen brush structure in an aqueous solution which reduces the interfacial free energy between the hydrated brush and footprint protein secreted from the paired attachment disks of the ambulatory antennules of cypris larvae. However, it is still unclear why polyzwitterion brushes result in a good antifouling performance rather than polyanion and polycation brushes, even though these surfaces show superhydrophilicity and wettability. As mentioned by Yang [29], the poly(HEMA) brush also showed excellent prevention of cyprids settlement, however, its surface is rather hydrophobic as compared to polyelectrolytes. Therefore, the surface hydrophilicity alone could not explain the fouling-resistance mechanism. Evaluation of the interaction forces between the substrate and antennules of cypris larvae is necessary to understand the attachment factors of cypris larvae. In the following section, the force measurement of cyprids and their footprint proteins is described.

3.4 Force Measurements Between Cypris and Substrates

The first trial of the direct measurement of the adhesion strength of live cypris larvae was performed by Yue and coworkers [10,11]. As shown in Figure 3.3, the authors attached live cypris larvae of Balanus balanoides to a piece of nichrome wire (d = 0.1 mm, L = 10 mm) with glue. The rectangular glass dish was filled with seawater and placed on a rack-controlled micrometer stage beneath the Cahn balance box. The wire was perpendicularly hooked on a suspension wire connected to a Cahn electro-microbalance to detect the force required to detach the cypris larvae from a slate substrate placed in seawater, by controlling the stage position with a micrometer. The observed separation force (N) of attached cypris from the slate was converted into adhesion strength ($N \cdot m^{-2}$) using the area of attachment of ambulatory antennules (tentacles). The slate substrate was treated with arthropodin (protein) solution prepared from *Balanus balanoides* adults. The cleaned slate with distilled water was also used for the control experiment.

The nichrome wire was immobilized on a cypris following two ways. In first strategy, the wire was parallel to the length of the cypris body so that cypris was vertically oriented when the tentacles attached to the slate. In the other case, the wire was glued in the direction vertical to the length of the cypris body so that cypris formed a transverse attachment when the slate was pulled. However, the authors reported no significant effect of the orientation of cypris attachment. They carried out a series of force measurements at intervals during a period of six weeks from April to May in 1979. It was found that the adhesion strength increased from 1×10^5 to 2×10^5 N·m^{-2} in

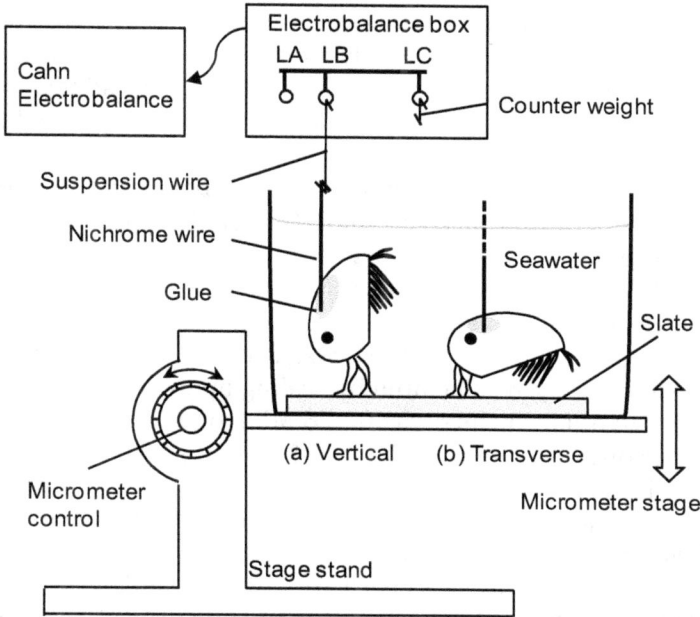

Figure 3.3 Schematic view of the apparatus (not to scale) used to measure the force required to detach cyprids of Balanus balanoides from the slate substrate pretreated with arthropodin in seawater, as reported by Yule and coworkers [10]. The nichrome wire (d = 0.1 mm, L = 10 mm) was glued with cyprids for (a) vertical and (b) transverse attachment. LA, LB, and LB in the electrobalance box are balance loops A, B, and C, respectively.

mid-April to early May, reaching a maximum in the range of 2.4×10^5 N·m^{-2} (= 0.24 MPa), and subsequently dropped sharply at the end of May. The seasonal variation in the temporary attachment of cypris was quantitatively evaluated. The authors also found that the force required to detach the cyprids from the arthropodin-coated slate was significantly larger than that from the cleaned slate in all experiments, indicating that the arthropodin protein induced a greater adhesion of the cyprids. These experiments exhibit a pioneering work for direct measurement of the adhesion strength of cypris larvae, however, the interaction forces between the substrate surface and the footprint protein secreted from the paired attachment disks of the ambulatory antennules of larvae were unclear.

The footprint deposition is the key for understanding the barnacle attachment. The footprint protein secreted from the paired attachment disks of the ambulatory antennules of larvae contains SIPC [35] to achieve temporary adhesion during surface exploration and acts as a pheromone [36] of cuticular glycoprotein [37], inducing the gregarious settlement of conspecific cyprids [38-39]. The adhesion force of the footprint protein was first evaluated by Vancso *et al.* [40,41] using AFM. The AFM cantilever was controlled to contact once the footprint proteins deposited by *Semibalanus balanoides* for temporary attachment during surface exploration on the glass surface modified with amino (NH_2)-terminated and methyl-terminated alkylsilane, followed by pulling off in seawater to obtain a force-distance curve. The observed interaction force between the footprint proteins and AFM tip made of Si_3N_4 was 0.41 nN. The estimated protein adhesion strength calculated by scaling up the force on the AFM tip to the size of foot prints was 0.026 MPa, which was much lower than the empirical value (0.068 - 0.076 MPa) reported by Yule and Walker [11]. Vancso *et al.* [40,41] also conducted the force-curve measurement of the footprints deposited by *Amphibalanus amphitrite* (*Balanus amphitrite*) cyprids on a chemically modified glass surface in seawater to obtain a characteristic sawtooth profile corresponding to the visco-adhesive mechanism of footprint proteins.

Vancso and co-workers further investigated the adhesion force of footprint proteins by AFM using a colloidal probe-based cantilever to demonstrate the suppression of protein adsorption on poly(sulfobetaine) brushes [42] and the effect of the substrate surface hydrophilicity on the interaction forces of the cyprid adhesive protein [12]. The authors prepared aldehyde-functionalized silica probe shown in Figure 3.4(a) to make contact with the footprint proteins of the cypris larvae of A. amphitrite. The footprint proteins were transferred to the probe surface by the chemical reaction of the aldehyde and the amino group of the protein. The authors carried out the force-curve measurement in autoclaved filtered seawater on the hydrophilic and hydrophobic substrates immobilized with the hydroxy group and pentafluorophenyl group, respectively, as shown in Figure 3.4(b)(c). The measurements using the surface-modified colloid probe bearing footprint proteins revealed a greater adhesion force (21 nN) on the hydrophobic surfaces compared with the more hydrophilic surfaces (7.2 nN). These results suggested that the adhesion strength was mainly driven by the hydrophobic interactions between the footprint proteins and substrate. However, the authors also described that the surface wettability or surface free energy alone is not the decisive factor determining the final settlement of cyprids [43,44]. Although they pointed out that the direct scaling of adhesion forces observed by force-curve measurements could not predict the real adhesion of cyprids, however, the AFM-based methodology is particularly useful for the quantitative evaluation or comparison of the molecular interaction of proteins.

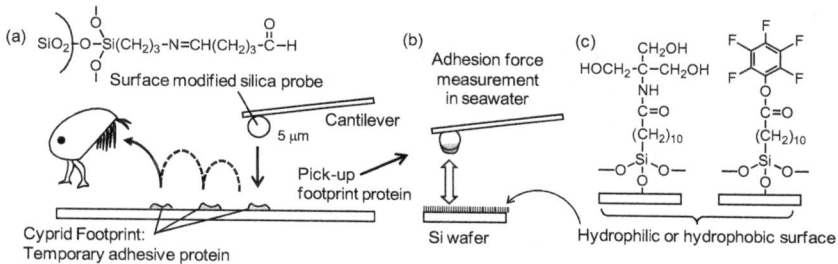

Figure 3.4 Schematic view of the adhesion force measurement using a chemically modified silica probe with cyprid footprint protein on a hydrophilic or hydrophobic surface, reported by Vancso *et al.* [12]: (a) a chemical structure of functional group on the silica, (b) preparation of silica probe bearing footprint protein and (c) chemical structure of functional group on the silicon wafers.

AFM technique was further used for the direct measurement of the interaction strength of live cyprids tentacles on the chemically modified glass surface [13]. Shiomoto and coworkers performed the piezoelectric scanner manipulation of AFM to capture a live cypris larva of *Megabalanus rosa* successfully in diluted seawater (salinity 22) by the cantilever head bearing an elastic-type chemically reactive adhesive consisting of 60% modified silicone polymer and 40% synthetic resin. The direction of the cantilever was perpendicular to the craniocaudal axis of the cypris larva, as shown in Figure 3.5. In fact, the unique behavior of the cypris larva was observed during the capture process. The settlement rate and activity of the cypris larvae of M. rosa were temporally reduced in diluted seawater (salinity 22) with a low salt concentration. These were promoted again in seawater with salinity 35 (normal salt concentration of seawater) [45]. Then, the seawater (salinity 22) was exchanged with filtered seawater (salinity 35) to commence the adhesion experiment. Two pieces of quartz cover glass, one grafted with a polymer brush and one coated with a propyltrimethoxysilane (PrS) monolayer, were aligned on the flat slide glass and stabilized with glue, as shown in Figure 3.5. Using the micrometer of the microscope moving stage, the cover glass was moved near the tentacles of the live cypris. Then, the cypris larva began tentatively touching the side face of the cover glass with its tentacles to explore the surface. During the touching and detaching of the tentacles, the cantilever twisted in accordance with the adhesion strength between the tentacles and substrate surface. The torsion of the cantilever was detected by the photodiode as a lateral deflection (V) of the laser and was converted to adhesion force (μN) by multiplying with the torsion spring constant [46] and cantilever sensitivity. An experiment was carried out in seawater (salinity 35) by repeating the detachment of cypris immobilized on the cantilever towards/from the cover glass.

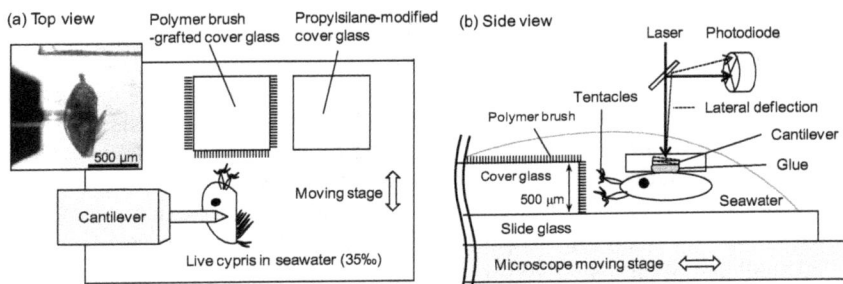

Figure 3.5 (a) Top view and (b) side view of the setting for adhesion force measurement of a live cypris Megabalanus rosa in seawater (salinity 35) on side face of a cover glass (thickness = 500 μm) modified with hydrophilic polymer brushes or hydrophobic propylsilane monolayer, reported by Shiomoto *et al.* [13]. Lateral deflection of AFM cantilever bearing a live cypris was monitored by a photodiode when the tentacles were attached on and detached from the cover glass, by moving the microscope stage.

The adhesion experiments with live cypris were carried out every two or three days after cyprids metamorphosed from nauplii, using two sets of quartz cover glass, one modified with a hydrophilic polymer brush and the other coated with a hydrophobic PrS monolayer, as shown in Figure 3.5, to compare the adhesion behavior of cypris on the two substrates under the same aqueous environment.

Young cypris exhibited within 10 days adhesion force lower than 4 μN on the hydrophobic PrS surface, but the adhesion force gradually increased with age and reached 30-40 μN after 14-20 days, as shown in Figure 3.6. These results indicated that the adhesive

behavior of cypris was activated with age until two weeks after transformation from nauplii. The adhesion force dropped again 22 days
later, probably due to deterioration with age. On the other hand, the
surfaces of the hydrophilic poly(MAPS) and poly(HEMA) brushes exhibited extremely low adhesion of cypris larva for 21 days in seawater (the results of the poly(HEMA) brush are not shown here).
These trends agreed with the previous results reported by Yang [29]
and Higaki *et al.* [30], describing the effective antifouling properties
of hydrophilic polymer brushes. In contrast, the MAPS monolayer
surface showed 30 μN of adhesion force with cypris after 14-20 days,
similar to a hydrophobic PrS-modified surface, as shown in Figure
3.6(b). The result suggested that tentacles attached to the surface
strongly, even if the sulfobetaine monolayer covered the substrate
surface. Therefore, low adhesion on the poly(MAPS) brush surface
might be induced not only by the poor chemical interaction between
MAPS and the footprint protein, but also by a swollen brush structure
in a salt aqueous solution due to the screening of electrostatic interaction between sulfobetains and exclusive volume effect of high graft
density [47-48].

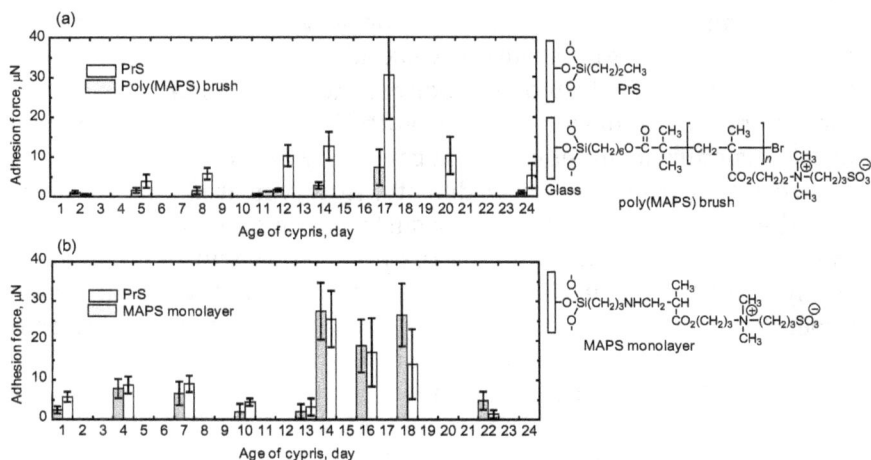

Figure 3.6 Age dependency of adhesion force of tentacles of a live cypris
on surface of (a) poly(MAPS) brush and (b) MAPS monolayer in seawater
(salinity 35) at 23 °C measured by AFM. As a control experiment, adhesion
forces on the surface of a hydrophobic propylsilane (PrS) monolayer were
measured for each force measurement (see Figure 3.5). The poly(HEMA)
brushed also showed low adhesion strength similar to the poly(MAPS)
brush.

Supposing that the typical footprint of one tentacle was approximately 20 μm in diameter, the adhesion strength required to detach the live cypris from the hydrophobic PrS surface was calculated to be $0.95 \times 10^5 - 1.9 \times 10^5$ N·m^{-2} (0.095-0.190 MPa), which is lower than the vertical force of 0.24 MPa of the cypris larvae of *B. Balanoides* for detachment from a slate substrate, as reported by Yule and Crisp [10]. Yule and Walker [11] also reported an adhesion strength of 0.069 - 0.076 MPa on a glass substrate. We could not directly compare the adhesion force of different species of cypris larvae on various substrate surfaces, however, it is considered meaningful that a similar order of magnitude of adhesion force was obtained from different measurement systems. Overall, AFM method is appropriate for measuring relatively weak adhesion or repeatable temporary adhesion varying with time of living entities, under various environmental conditions.

3.5 Conclusions

Ion-containing polymer brushes, in particular polyzwitterion brushes, prepared by the "grafting-from" method using ATRP exhibit excellent antifouling characteristics for preventing the settlement of cypris larvae, due to extremely weak adhesive interaction between the tentacles and hydrated polymer brushes. Additional possible factor for the low adhesion on polyzwitterion brushes is the swollen brush structure in aqueous media caused by an exclusive volume effect resulting from high graft density. However, it should be noted that the surface hydrophilicity or surface free energy alone are not the decisive factors determining the final settlement of cyprids. Recently, the effective prevention of cyprid settlement was achieved by hyperbranched polymers consisting of the well-designed mixture of hydrophobic and hydrophilic polymer segments [49-50]. This chapter also described the measurements of the adhesion force of live cypris in seawater by AFM technique. A live cypris was successfully immobilized on the cantilever to make contact with the chemically modified surfaces. The age dependence of the adhesion behavior of the cypris was clearly observed. It has already been reported that the settlement activity of cypris is largely influenced not only by age, but also by season, environment and water flow rate (stream velocity) near the substrates [51]. Further experiments should focus on the adhesion measurements of not only cypris larvae, but also various sessile animals, under different conditions.

References

1. Callow, J. A., and Callow, M. E. (2011) Trends in the development of environ-mentally friendly fouling-resistant marine coatings. *Nature Communications*, **2**, 244.
2. Yang, W. J., Neoh, K. G., Kang, R. T., Teo, S. L. M, and Rittschof, D. (2014) Polymer brush coatings for combating marine biofouling. *Progress in Polymer Science*, **39**, 1017-1042.
3. Lejars, M., Margaillan, A., and Bressy, C. (2012) Fouling release coatings: a nontoxic alternative to biocidal antifouling coatings. *Chemical Reviews*, **112**, 4347-4390.
4. Clare, A. S., and Matsumura, K. (2000) Nature and perception of barnacle settlement pheromones. *Biofouling*, **15**, 57-71.
5. Walker, G., and Yule, A. B. (1984) Temporary adhesion of the barnacle cyprid: the existence of an antennular adhesive secretion. *Journal of the Marine Biological Association of the United Kingdom*, **64**, 679-686.
6. Dreanno, C., Kirby, R. R., and Clare, A. S. (2006) Smelly feet are not always a bad thing: the relationship between cyprid footprint protein and the barnacle settlement pheromone. *Biology Letters*, **2**, 423-425.
7. Crisp, D. J., and Meadows, P. S. (1963) Adsorbed layers: the stimulus to settlement in barnacles. *Proceedings of the Royal Society of London. Series B: Biological Sciences*, **158**, 364-87.
8. Rühe, J. (2004) Polymer brushes: On the way to tailor-made surfaces. In: Polymer Brushes: Synthesis, Characterization and Applications, Advincula, R. C., Brittain, W. J., Caster, K. C., and Rühe, J. (eds.), Wiley-VCH, Germany, pp. 1-31.
9. Tsujii, Y, Ohno, K, Yamamoto, S, Goto, A, and Fukuda, T. (2006) Structure and properties of high-density polymer brushes prepared by surface-initiated living radical polymerization. *Advances in Polymer Science*, **197**, 1-45.
10. Yule, A. B., and Crisp, D. J. (1983) Adhesion of cypris larvae of the barnacle, Balanus Balanoides, to clean and arthropodin treated surfaces. *Journal of the Marine Biological Association of the United Kingdom*, **63**, 261-271.
11. Yule, A. B., and Walker, G. (1984) The temporary adhesion of barnacle cyprids: Effects of some differing surface characteristics. *Journal of the Marine Biological Association of the United Kingdom*, **64**, 429-439.
12. Guo, S., Puniredd, S. R., Jańczewski, D., Lee, S. S. C., Teo, S. L. M., He, T., Zhu, X., and Vancso, G. J. (2014) Barnacle larvae exploring surfaces with variable hydrophilicity: Influence of morphology and adhesion of 'footprint' proteins by AFM. *ACS Applied Materials & Interfaces*, **6**, 13667-13676.

13. Shiomoto, S., Yamaguchi, Y., Yamaguchi, K., Nogata, Y., and Kobayashi, M. (2019) Adhesion force mesurement of live cypris tentacles by scanning probe microscopy in seawater. *Polymer Journal*, **51**, 51-59.

14. Yang, W. J., Cai, T., Neoh, K. G., and Kang, E. T. (2011) Biomimetic anchors for antifouling and antibacterial polymer brushes on stainless steel. *Langmuir*, **27**, 7065-7076.

15. Almeida, E., Diamantino, T. C., and de Sousa, O. (2007) Marine paints: the particular case of antifouling paints. *Progress in Organic Coatings*, **59**, 2-20.

16. Patwardhan, S. V., Taori, V. P., Hassan, M., Agashe, N. R., Franklin, J. E., Beaucage, G., Mark, J. E., and Clarson, S. J. (2006) An investigation of the properties of poly(dimethylsiloxane)- bioinspired silica hybrids. *European Polymer Journal*, **42**, 167-178.

17. Stein, J., Truby, K., Wood, C. D., Stein, J., Gardner, M., Swain, G., Kavanagh, C., Kovach, B., Schultz, M., Wiebe, D., Holm, E., Montemarano, J., Wendt, D., Smith, C., and Meyer, A. (2003) Silicone foul release coatings: effect of the interaction of oil and coating functionalities on the magnitude of macrofouling attachment strengths. *Biofouling*, **19**, 71-82.

18. Lee, S. B., Koepsel, R. R., Morley, S. W., Matyjaszewski, K., Sun, Y. J., and Russell, A. J. (2004) Permanent, nonleaching antibacterial surfaces. 1. Synthesis by atom transfer radical polymerization. *Biomacromolecules*, **5**, 877-882.

19. Palermo, E. F, Vemparala, S., and Kuroda, K. (2012) Cationic spacer arm design strategy for control of antimicrobial activity and conformation of amphiphilic methacrylate random copolymers. *Biomacromolecules*, **13**, 1632-1641.

20. Sugiyama, K., and Aoki, H. (1994) Surface modified polymer microspheres obtained by the emulsion copolymerization of 2-methacryloyloxyethyl phosphorylcholine with various vinyl monomers. *Polymer Journal*, **26**, 561-569.

21. Ishihara, K. (2000) Bioinspired phospholipid polymer biomaterials for making high performance artificial organs. *Science and Technology of Advanced Materials*, **1**, 131-138.

22. Ishihara, K., Ueda, T., and Nakabayashi, N. (1990) Preparation of phospholipid polymers and their properties as polymer hydrogel membranes. *Polymer Journal*, **22**, 355-360.

23. Iwata, R., Suk-In, P., Hoven, V. P., Takahara, A., Akiyoshi, K., and Iwasak, Y. (2004) Control of nanobiointerfaces generated from well-defined, biomimetic polymer brushes for protein and cell manipulations. *Biomacromolecules*, **5**, 2308-2314.

24. Kobayashi, M., Terayama, Y., Yamaguchi, H., Terada, M., Murakami, D., Ishihara, K., and Takahara, A. (2012) Wettability and antifouling behavior on the surfaces of super hydrophilic polymer brushes.

Langmuir, **28**, 7212-7222.

25. Higaki, Y., Kobayashi, M., Murakami, D., and Takahara, A. (2016) Anti-fouling behavior of polymer brush immobilized surfaces. *Polymer Journal*, **48**, 325-31.

26. Ladd, J., Zhang, Z., Chen, S., Hower, J. C., Jiang, S., and Zwitterionic (2008) Polymers exhibiting high resistance to nonspecific protein adsorption from human serum and plasma. *Biomacromolecules*, **9**, 1357-1361.

27. Kitano, H., Suzuki, H., Kondo, T., Sasaki, K., Iwanaga, S., Nakamura, M., Ohno, K., and Saruwatari, Y. (2011) Image printing on the surface of anti-biofouling zwitterionic polymer brushes by ion beam irradiation. *Macromolecular Bioscience*, **11**, 557-564.

28. Aldred, N., Li, G. Z., Gao, Y., Clare, A. S., and Jiang, S. Y. (2010) Modulation of barnacle (Balanus amphitrite Darwin) cyprid settlement behavior by sulfobetaine and carboxybetaine methacrylate polymer coatings. *Biofouling*, **26**, 673-683.

29. Yang, W. J., Neoh, K.G., Kang, E. T., Lee, S. S. C., Teo, S. L. M., and Rittschof, D. (2012) Functional polymer brushes via surface-initiated atom transfer radical graft polymerization for combating marine biofouling. *Biofouling*, **28**, 895-912.

30. Higaki, Y., Nishida, J., Takenaka, A., Yoshimatsu, R., Kobayashi, M., and Takahara, A. (2015) Versatile inhibition of marine organism settlement by zwitterionic polymer brushes. *Polymer Journal*, **47**, 811-818.

31. Murata, H., Koepsel, R. R., Matyjaszewski, K., and Russell, A. J. (2007) Permanent, non-leaching antibacterial surfaces. 2: How high density cationic surfaces kill bacterial cells. *Biomaterials*, **28**, 4870-4879.

32. Matsugi, T., Saito, J., Kawahara, N., Matuso, S., Kaneko, H., Kashiwa, N., Kobayashi, M., and Takahara, A. (2009) Surface modification of polypropylene molded sheets by means of surface-initiated ATRP of methacrylates, *Polymer Journal*, **41**, 547-554.

33. Yuan, S. J., Pehkonen, S. O., Ting, Y. P., Neoh, K. G., and Kang, E. T. (2010) Antibacterial inorganic-organic hybrid coatings on stainless steel via consecutive surface-initiated atom transfer radical polymerization forbiocorrosion prevention. *Langmuir*, **26**, 6728-6736.

34. Yuan, S. J., Pehkonen, S. O., Ting, Y. P., Neoh, K. G., and Kang, E. T. (2009) Inorganic-organic hybrid coatings on stainless steel by layer-by-layer deposition and surface-initiated atom-transfer-radical-polymerization for combating biocorrosion. *ACS Applied Materials and Interfaces*, **1**, 640-652.

35. Matsumura, K., Nagano, M., and Fusetani, N. (1998) Purification of a settlement-inducing protein complex (SIPC) of the barnacle, Balanus amphitrite. *Journal of Experimental Zoology*, **281**, 12-20.

36. Yule, A. B., and Walker, G. (1985) Settlement of balanus balanoides: the effect of cyprid antennular secretion. *Journal of the Marine Biological Association of the United Kingdom*, **65**, 707-712.
37. Dreanno, C., Kirby, R. R., and Clare, A. S. (2006) Locating the barnacle settlement pheromone: spatial and ontogenetic expression of the settlement-inducing protein complex of Balanus amphitrite. *Proceedings of the Royal Society B: Biological Sciences*, **273**, 2721-2728.
38. Clare, A. S., Freet, R. K., and McClary, M. (1994) On the antennular secretion of the cyprid of Balanus Amphitrite, and its role as a settlement pheromone. *Journal of the Marine Biological Association of the United Kingdom*, **74**, 243-250.
39. Clare, A. S., and Matsumura, K. (2000) Nature and perception of barnacle settlement pheromones. *Biofouling*, **15**, 57-71.
40. Phang, I. Y., Aldred, N., Clare, A. S., and Vancso, G. J. (2008) Towards a nanomechanical basis for temporary adhesion in barnacle cyprids (Semibalanus Balanoides). *Journal of the Royal Society Interface*, **5**, 397-401.
41. Phang, I. Y., Aldred, N., Ling, X. Y., Huskens, J., Clare, A. S., and Vancso, G. J. (2010) Atomic force microscopy of the morphology and mechanical behaviour of barnacle cyprid footprint proteins at the nanoscale. *Journal of the Royal Society Interface*, **7**, 285-296.
42. Schon, P., Kutnyanszky, E., ten Donkelaar, B., Santonicola, M. G., Tecim, T., Aldred, N., Clare, A. S., and Vancso, G. J. (2013) Probing biofouling resistant polymer brush surfaces by atomic force microscopy based force spectroscopy. *Colloids and Surfaces B: Biointerfaces*, **102**, 923-930.
43. Petrone, L., Di Fino, A., Aldred, N., Sukkaew, P., Ederth, T., Clare, A. S., and Liedberg, B. (2011) Effects of surface charge and Gibbs surface energy on the settlement behaviour of barnacle cyprids (Balanus amphitrite). *Biofouling*, **27**, 1043-1055.
44. Di Fino, A., Petrone, L., Aldred, N., Ederth, T., Liedberg, B., and Clare, A. S. (2014) Correlation between surface chemistry and settlement behaviour in barnacle cyprids (Balanus improvisus). *Biofouling*, **30**, 143-152.
45. Nogata, Y., Tokikuni, N., Yoshimura, E., Sato, K., Endo, N., Matsumura, K., and Sugita, H. (2011) Salinity limitations in larval settlement of four barnacle species. *Sessile Organisms*, **28**, 47-54.
46. Mullin, N., and Hobbs, J. K. (2014) A non-contact, thermal noise based method for the calibration of lateral deflection sensitivity in atomic force microscopy. *The Review of Scientific Instruments*, **85**, 113703.
47. Kikuchi, M., Terayama, Y., Ishikawa, T., Hoshino, T., Kobayashi, M., Ogawa, H., Masunaga, H., Koike, J., Horigome, M., Ishihara, K., and Takahara, A. (2012) Chain dimension of polyampholytes in solution

and immobilized brush states. *Polymer Journal*, **44**, 121-130.

48. Higaki, Y., Inutsuka, Y., Sakamaki, T., Terayama, Y., Takenaka, A., Higaki, K., Yamada, N. L., Moriwaki, T., Ikemoto, Y., and Takahara, A. (2017) Effect of charged group spacer length on hydration state in zwitterionic poly(sulfobetaine) brushes. *Langmuir*. **33**, 8404-8412.

49. Imbesi, P. M., Finlay, J. A., Aldred, N., Eller, M. J., Felder, S. E., Pollack, K. A., Lonnecker, A. T., Raymond, J. E., Mackay, M. E., Schweikert, E. A., Clare, A. S., Callow, J. A., Callow, M. E., and Wooley, K. L. (2012) Targeted surface nanocomplexity: two-dimensional control over the composition, physical properties and anti-biofouling performance of hyperbranched fluoropolymer-poly(ethylene glycol) amphiphilic crosslinked networks. *Polymer Chemistry*, **3**, 3121-3131.

50. Imbesi, P. M., Gohad, N. V., Eller, M. J., Orihuela, B., Rittschof, D., Schweikert, E. A., Mount, A. S., and Wooley, K. L. (2012) Noradrenaline-functionalized hyperbranched fluoropolymer-poly(ethylene glycol) cross-linked networks as dual-mode, anti-biofouling coatings. *ACS Nano*, **6**, 1503-1512.

51. Larsson, A. I., Granhag, L. M., and Jonsson, P. R. (2016) Instantaneous flow structures and opportunities for larval settlement: Barnacle larvae swim to settle. *PLoS One*, **11**, 0158957.

4

Hybrid Organosilicone Materials as Efficient Anticorrosive Coatings in Marine Environment

Rami Suleiman,[a],* Amjad Khalil,[b] Mazen Khaled[c] and Bassam El Ali[c]

[a]Center of Research Excellence in Corrosion, King Fahd University of Petroleum & Minerals (KFUPM), Dhahran 31261, Saudi Arabia
[b]Department of Life Sciences, King Fahd University of Petroleum & Minerals (KFUPM), Dhahran 31261, Saudi Arabia
[c]Chemistry Department, King Fahd University of Petroleum & Minerals (KFUPM), Dhahran 31261, Saudi Arabia

*Corresponding author: ramismob@kfupm.edu.sa

4.1 Introduction

4.1.1 Corrosion Expenses

The deterioration of metallic infrastructure exposed to the marine environment due to corrosion and fouling phenomena is a serious challenge for various industries, especially the gas and oil industry. The worldwide cost of corrosion and related expenses in oil industries and applied companies, marine and aviation sectors, construction and even domestic life are estimated in trillions of US dollars (Figure 4.1) [1]. It is expected for these figures to increase in the future as the need for using metal in the infrastructure of the aforementioned industries is increasing day by day.

In the maritime environments, the presence of water, dissolved salts and humidity accelerates the corrosion of metallic infrastructures. As an example, the corrosion phenomenon is a significant problem in the Arabian Peninsula region due to the region's climate and enormous oil and gas activities as well as associated pollution factors. The structural components of the oil and gas industry in this region are suffering from the degradation of their engineering materials that are under continuous exposure to high levels of aggressive environmental parameters such as temperature, humidity and dusty winds.

Marine Coatings and Membranes, edited by Vikas Mittal
© 2019 Central West Publishing, Australia

Figure 4.1 Corrosion of immersed metallic panels in seawater environment.

Moreover, the biofouling of immersed metallic infrastructures (like ships) is another serious problem for marine industries. The growth of undesirable colonies of microorganisms (algae and bacteria) and macro niches (like barnacles and marine weeds) on vessels result in the wastage of fuel, have significant environmental implications, along with economic consequences (Figure 4.2). Problems and consequences of material failure through corrosion and biofouling could range from minor to severely catastrophic, with loss of lives and disruption of essential services, if the mechanism of the process (and not just the causes) is not promptly addressed [2]. It is estimated

Figure 4.2 Accumulation of foulants on the surface of metallic panels exposed to marine environment.

that the annual corrosion and fouling costs of the US marine shipping industry was at US $ 2.7 billion in 1998, and this number is increasing with time [3,4].

4.1.2 Anticorrosion/Antifouling Coatings

The application of protective coatings is by far the most generic way to protect metallic infrastructures against corrosion and fouling. As an example, and among the various known corrosion control technologies, the use of coatings was the dominant technology to control corrosion in the upstream oil operations sector of ARMACO company in Saudi Arabia during the period 1998 to 2001 (Figure 4.3) [5].

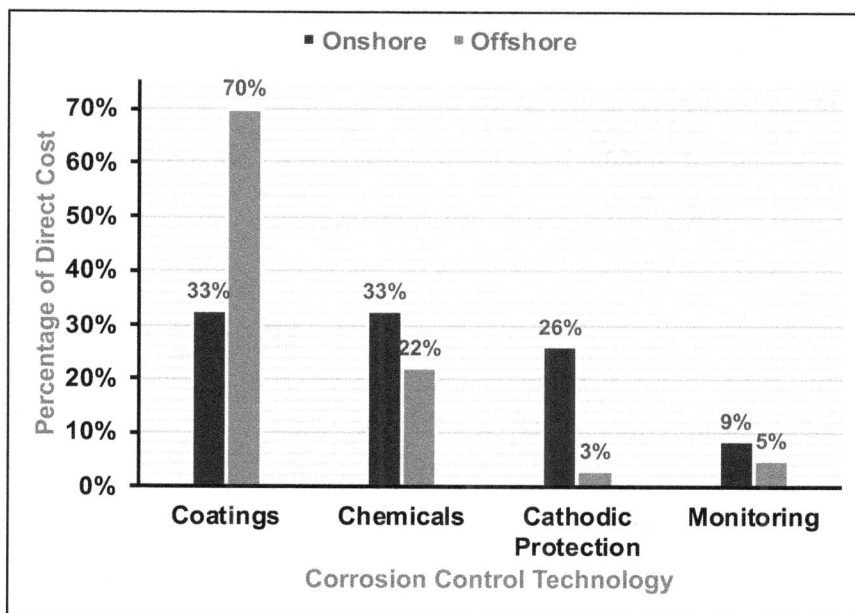

Figure 4.3 Percentage of direct cost for specific corrosion control technologies, onshore (blue (first column), 2000-2001) and offshore (red (second column), 1998-2001) productions (percentage values were adapted from [3]).

Protective anticorrosive coatings can be classified as per the mechanism by which they protect the metal surface against corrosion, i.e. barrier, sacrificial and inhibitive protection effects [6]. The chemistry of the applied coating should be carefully optimized so as to maximize the desired properties of the deposited coating on metal surface such as barrier, thermal, hardness, morphological and adhesion properties [7,8]. The type of metal substrate, deposition technique, curing time, corrosive environment and many other parameters should be taken into consideration during the selection/development of a protective coating. Human and environmental concerns which are attributed to the use of volatile organic compounds (VOC) or toxic additives such as chromium in the coating formulations should be also carefully recognized by coatings' fabricators and end-users. Coatings usually protect metal surfaces against corrosion by inhibiting the passage of corrosive ions such as H^+ or Cl^- to reach the metal surface. This transport process of aggressive species is mainly dependent on the physical and chemical properties of the coating and the substrate surface, and any inhomogeneities in the coating resulting from the presence of air bubbles, cracks, microvoids, contaminants, trapped solvents, nonbonded and weak areas, pigment-resins and coating-substrate interfacial layers ultimately affect the degradation process of metals [6]. Various additives can be added in small amounts to the coating formulation which serve a specific function. These additives can be binder, extender, curing agent, inhibitor, antibacterial agent, stabilizer and solvents.

Metal coated structures can be exposed to different corrosive environments such as immersion in seawater or acid, burial in soil, atmospheric pollution and ultraviolet radiation. The various components of these corrosive environments determine directly the specific requirements of the protective coating and its application methodology as well as exposure time. However, it is very difficult to rank or give a specific corrosivity index to a corrosive environment as the latter is affected by different parameters such as temperature, salts, bacteria, humidity, dissolved oxygen, etc. The marine environment affects the metallic structures that are either exposed to the marine atmosphere or those immersed under seawater. In contrast to fresh water, seawater environment contains high amounts of the aggressive chloride ions which may cause pitting corrosion [9].

The protective coating systems usually involve the application of primer, middle and top coats on top of each other (triple coat system) and on the metal surface. Every coating 'layer' in any conventional protective system has a specific function, and the different layers should be applied in a particular sequence. Primers are applied mainly on the metal surfaces to protect them from corrosion and to enhance the adhesion properties of the coating layers. The middle (intermediate) coat provides the required thickness to the coating system and help in promoting its barrier properties, where as the top coat provides one or more of the required specific functionalities to the coating system such as hydrophobicity, antifouling, color and UV-radiation protection. It is worth mentioning here that the new trend in the coating research is to develop a single coat system in which the coating layer has the combined properties of the three coating layers in the tri-coat system (Figure 4.4). The current protective coating systems for immersed metallic structures in seawater normally include an anticorrosive primer and an antifouling topcoat. In some cases, where the anticorrosive primer may contain some components that reduce the adhesion properties of the antifouling paint, an intermediate tie coat can be applied between the primer and the antifouling paint [10].

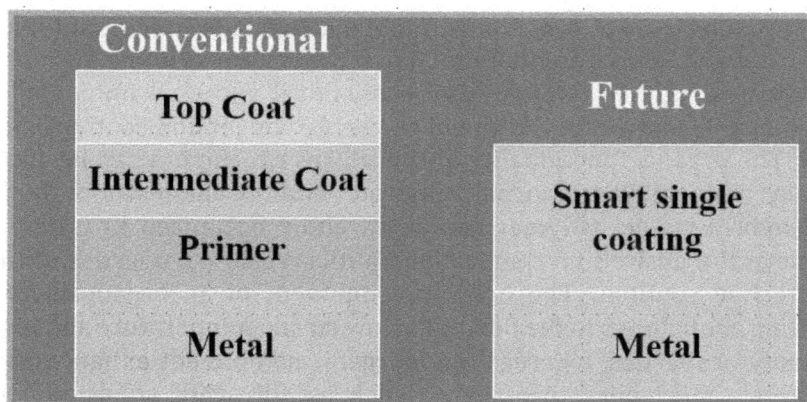

Figure 4.4 Conventional versus future anticorrosive coating systems.

Depending on their base chemical structures, anticorrosive coatings can be metallic, non-metallic (organic) or hybrid polymeric materials. Metallic coatings compose of a metallic element or alloy such as chromium, nickel, copper and cadmium, and work mainly on creating a layer that modifies the properties of the workpiece surface compared to those of the metal being applied [11]. Non-metallic coatings refer to the class of naturally-occurring or synthetic polymeric materials that are electrically non-conductive, relatively inert and do not have a metal in their base polymeric network such as paints, lacquers, oils and waxes, etc. These coatings, when uniformly applied, provide a continuous protection to the metal surface as long as the surface film is not damaged [12]. Hybrid coatings refer to the polymeric materials that have organic components attached to an inorganic polymeric network. These materials have received a significant attention from the coating developers as these combine the preferred properties of organic polymers (e.g., flexibility, dielectric, ductility) and inorganic materials such as silica (e.g., high thermal strength, stability, UV-vis absorbance, hardness), leading to a promising application as protective coatings for various metallic substrates [13-15]. Organosilicones are a famous example of hybrid polymer materials and more discussion on the chemistry, properties and applications of these materials as protective coatings is presented below.

4.1.3 Hybrid Organosilicone Coatings

Hybrid organic-inorganic sol-gel coatings prepared via the hydrolysis/condensation of organofunctional alkoxysilane precursors have been developed as potential functional protection systems for metal substrates, as they offer a more economical and environmentally-friendly alternative to traditional chromate conversion coatings [16-18]. The organic functionality of the silanes precursors can be alkyl, epoxy, aminoalkyl, cyano, haloalkyl, etc., and the chemistry of the final hybrid sol-gel polymer can be carefully optimized in order to achieve the desired mechanical and anticorrosion properties of the protective coatings. The organic component of an organosilicone coating contributes to the flexibility, low curing temperature and high density properties, whereas the inorganic component enhances the coating's mechanical properties, such as substrate adhesion and hardness (Figure 4.5) [7].

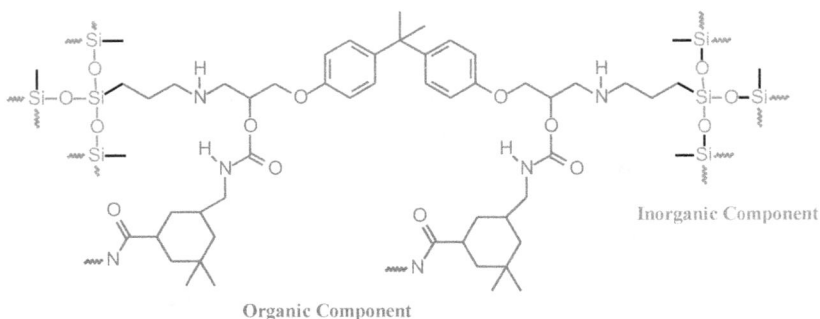

Figure 4.5 Illustration of the organic and inorganic components in the chemical structure of the hybrid sol-gel materials, reported in Reference 7.

In addition to the methodology of synthesizing the hybrid sol-gel coatings via the hydrolysis/polycondensation of the organofunctional alkoxysilane precursors (Figure 4.6), these materials can also be prepared by the blending of epoxy precursors/resins with polysiloxanes/organosiloxanes and difunctional aminosilane hardeners [16,19,20].

The availability of various organosilane precursors to generate hybrid sol-gel coatings, their scalability and ease of application, and the ability to control the preparation parameters make sol-gel methodology a unique process for the preparation of this class of protective coatings [21].

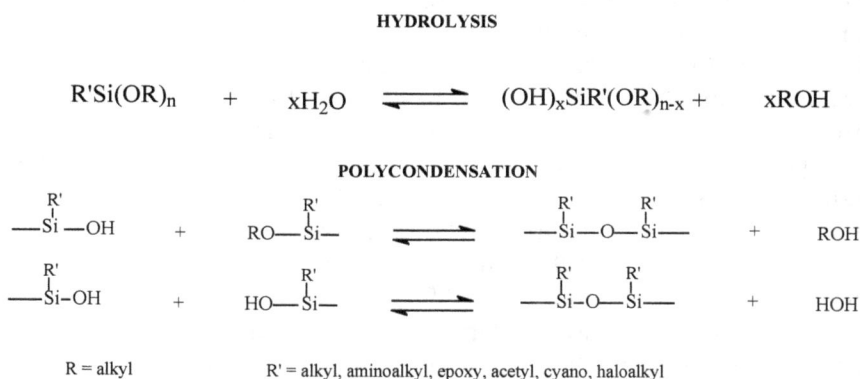

HYDROLYSIS

$$R'Si(OR)_n \quad + \quad xH_2O \quad \rightleftharpoons \quad (OH)_xSiR'(OR)_{n-x} + \quad xROH$$

POLYCONDENSATION

$$\underset{\underset{R'}{|}}{-Si}-OH \quad + \quad RO-\underset{\underset{R'}{|}}{Si}- \quad \rightleftharpoons \quad -\underset{\underset{R'}{|}}{Si}-O-\underset{\underset{R'}{|}}{Si}- \quad + \quad ROH$$

$$\underset{\underset{R'}{|}}{-Si}-OH \quad + \quad HO-\underset{\underset{R'}{|}}{Si}- \quad \rightleftharpoons \quad -\underset{\underset{R'}{|}}{Si}-O-\underset{\underset{R'}{|}}{Si}- \quad + \quad HOH$$

R = alkyl   R' = alkyl, aminoalkyl, epoxy, acetyl, cyano, haloalkyl

Figure 4.6 Chemical reactions involved in the preparation of hybrid sol-gel materials.

Recently, many research studies have reported that sol-gel coatings can be successfully functionalized with certain additives such as inhibitors, antibacterial agents and hardeners, thus, yielding improved multifunctional properties of the base sol-gel coating [8,22,23]. The success of the functionalization process of the parent hybrid sol-gel matrix requires a careful selection of the type of embedded additive and the reaction conditions. Eduok *et al.* [24] have reviewed the chemistry and physical properties of various sol-gel coating formulations applied on metal substrates under various corrosive media [24]. As an example, in a study by Pérez *et al.* [25], the anticorrosion properties of a hybrid sol-gel coating applied on AA2024-T3 substrate were found to be significantly enhanced upon the addition of calcined red mud powder. The barrier properties for a hybrid sol-gel film on galvanized steel sheets were also found to be significantly improved in 0.1 M NaCl solution by the addition of montmorillonite clay nanoparticles. Previous studies by Suleiman *et al.* [27,28] have shown that corrosion inhibitors zinc aluminum polyphosphate (ZAPP) and molybdenum zinc oxide enhanced the ability of sol-gel coatings to protect mild steel from corrosion in 3.5 wt% NaCl. Corrosion inhibitors are generally added to a paint to increase its barrier properties and to help prevent surface disbonding. Disbonding may result from mechanical damage, thereby, creating cathodic sites, where the formation of hydroxyl ions leads to hydrolysis and oxidation reactions, thus, reducing coating adhesion. The purpose of corrosion inhibitors is to slow down the rate of the anodic reaction, and hence the cathodic disbondment reaction, by either decreased diffusion through the coating, surface adsorption of the inhibitor and/or the formation of a passive film on the surface of the metal. For example, ZAPP added to a sol-gel formulation was found to enhance pore resistance on immersion in Harrison's solution and delayed the formation of anodic sites [29]. A very recent study of Suleiman *et al.* [30] has proved that the addition of certain metal oxides (such as calcium, zinc and molybdenum oxides) to a parent hybrid sol-gel coating can greatly enhance its anticorrosion properties for mild steel samples immersed in 3.5% NaCl medium.

For marine applications, various research approaches have been attempted in order to replace toxic copper and TBT-based coatings used nowadays to mitigate the fouling of underwater metal surfaces. In one approach, it was proposed that the antifouling functionality can be achieved by modifying the method of preparation with compounds that can deter the adhesion of fouling organisms [10]. Another approach suggested the use of human- and environment-friendly coatings such as controlled depletion paints (CDPs) [31], tin-free self-polishing paints (TF-SPCs) [32] and biocide-free hybrid systems [33]. Recent studies have also shown that the functionality of the sol-gel matrix can be further modified to improve its antifouling properties by immobilized protective bacterium within its bulk [34]. For example, endospore-forming *Paenibacillus polymyxa* within a sol-gel coating has been shown to improve corrosion protection in the laboratory and at field trial locations [35,36]. More generally, a range of bacteria are known to produce compounds with antifouling properties [37], which can potentially be exploited if the bacteria are combined in the coating. *P. polymyxa* was selected for incorporation into the sol-gel matrix since endospores are able to withstand environments that other vegetative bacterial cells may not survive, such as low pH and high concentrations of solvent [35]. This bacterium is known for producing an antimicrobial compound (polymyxin) that aids the inhibition of the growth of microorganisms involved in corrosion and fouling [38]. Moreover, another *P. polymyxa* strain isolated from mollusc shells has previously been found to produce low molecular-mass surface active substances and to have an inhibitory effect on the growth of indicator micro-organsims including *Aspergillus niger*, *Candida albicans*, *Staphylococcus aureus* and *Bacillus subtilis* [39]. Other endospore-formers have also exhibited potential as antifouling agents. For example, an isolate of *B. licheniformis* has been reported to produce lipopeptide bio-surfactant with antimicrobial effects against a range of gram-positive and gram-negative bacteria [40]. Another example of natural antifouling activity from a endospore-former is the polysaccharide-producing *B. licheniformis* strain, isolated from the orange sponge *Spongia officinalis*, collected from a depth of 10 m in Sicilia, Italy. This bacterium was found to have anti-biofilm activity against a number of bacteria without having bactericidal effects [41]. The authors' approach to exploit natural anti-bio-fouling systems has been to immobilize micro-organisms with anti-fouling activities within a sol-gel coating [42]. As a continuation of the research efforts for the development of anticorrosion/antifouling hybrid sol-gel coatings, the case study discussed below aims to investigate the individual and synergistic effects of adding bacterial endospores and corrosion inhibitor on the anticorrosion and antifouling properties of a sol-gel coating doped with the two additives and applied on mild steel substrate. Field testing on control samples free from bacterial endospores (abiotic) has been also performed. The on-field fouling testing for all coating matrices, with and without the immobilized bacterium, is complemented with electrochemical studies and surface analytical evaluations.

4.2 Case Study

In this study, a new innovative hybrid sol-gel coating system has been synthesized, which was applied on mild steel panels and subsequently exposed to laboratory saline and field seawater environments in order to evaluate anticorrosion and antifouling properties. The hybrid coating was functionalized with zinc aluminum polyphosphate (ZAPP) and viable endospores of *Bacillus licheniformis* isolate (B6), all immobilized within the bulk of the coating to enhance anti-corrosion properties as well as fouling-resistance in seawater. The corrosion protection properties of all coating matrices applied to steel have been evaluated using electrochemical techniques as well as scanning electron microscopy (SEM). After immersion in open seawater for 6 weeks, the coating containing B6 bacterial strain endospores alone showed superior corrosion and antifouling protection compared to the coating containing ZAPP alone and the coating with both ZAPP and endospores. The field trial results also suggest that this class of sol-gel coating is incompatible with the ZAPP inhibitor.

4.2.1 Experimental Methods

Sol-gel Formulation Methods

The sol-gel coating used in this study was synthesized by blending 10 ml TEOS and 6 ml 3-glycidyloxypropyl-trimethoxysilane (GLYMO) for 2 h before adding 5 ml dodecylmethylsiloxane-hydroxypolyalkylene-oxypropyl methylsiloxane copolymer (dodecyl copolymer). The mixture was left to stir for another 2 h before adding 10 ml of a mixture of 0.05N HNO_3 and isopropyl alcohol (in a volume ratio of 1:2). The coating was aged for 48 h before use and labelled as abiotic "SG"; this solution had a pH of 5.8. 5 ml of the prepared sol-gel solution was then doped with Heucophos ZAPP® (ZAPP) corrosion inhibitor (5% w/v), sonicated for 15 minutes and stirred at 750 rpm for 24 h (and labelled as abiotic "SG-IZ"). A sol-gel bacterial/endospore suspension was then added to 10 mL of SG and SG-IZ mixtures and labelled as biotic "SG-B6" and "SG-IZB6", respectively.

Bacterial Strain

The protective bacterium used in this study is a new thermophilic *Bacillus licheniformis* strain 6 (B6) isolate from the hot springs of the Gazan area (Lat. 43°15E', Long. 16°56N', Southern Province of Saudi Arabia). The procedure for its isolation as well as genotypic and phenotypic characterization have been reported in detail elsewhere [43,44].

Bacterial Endospores Preparation

The bacterial endospore suspension has been prepared, characterized and mixed with the base hybrid sol-gel coating as per the procedure reported previously [34,45].

Substrate Preparation and Coatings' Application

Sol-gel coating samples were applied on steel panels using a K101 rod coating applicator (R K Print-Coat Instruments Ltd., UK), and subsequently cured at room temperature for 2 days under a laminar flow hood. The thickness of applied coatings was controlled by using a roller bar of defined wound wire size. Electrochemical masks (10 cm^2 from Gamry, US) were used to define the test area on the coated samples. The summary of the coating matrices (inhibitor and endospore-containing coatings) investigated in this study is given in Figure 4.7.

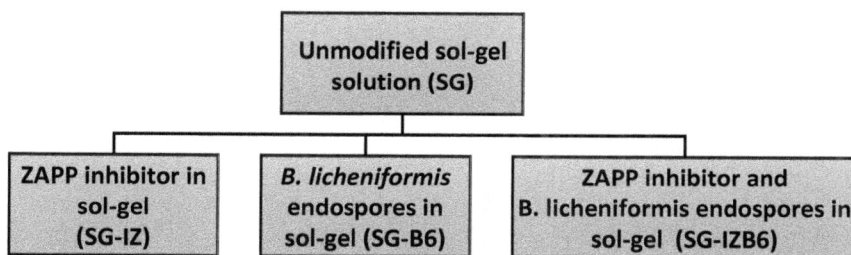

Figure 4.7 Summary of the coating (matrices inhibitor and endospore-containing coatings) investigated in this study.

Surface Morphology Characterization of the Coated Samples

The morphology of the coatings and cross-sections after corrosion testing were studied using a JEOL JSM6610LV SEM. Contact angle measurements were carried out on a DSA30 (KRUSS, Deutsch, Germany) instrument using a drop of deionized water. Aiming more accuracy, the measurements of the water drop contact angle were performed six times on each sample and the mean value of the six replicates is reported.

Confocal Microscopy

The confocal scanning laser microscopy (CLSM) characterization of the coated samples was performed as per our previous literature study [34]. This type of testing is very essential to prove endospore viability and to obtain insight on the distribution of vegetative bacterial cells within the base hybrid coating.

Field Trial at Half Moon Bay (KFUPM Beach) Saudi Arabia

The field testing of the anticorrosion and antifouling behavior of the coating matrices was conducted by immersing the coated panels under seawater of KFUPM beach, Half-Moon Bay, Al Khobar, East Province, Saudi Arabia (Lat. 26°28', Long. 50°20', Figure 4.8). The average seawater temperature during the study was 28 °C. The height of water in the test area was 1.5 m and the flow of water was minimal. The immersion period was between May and September of 2014.

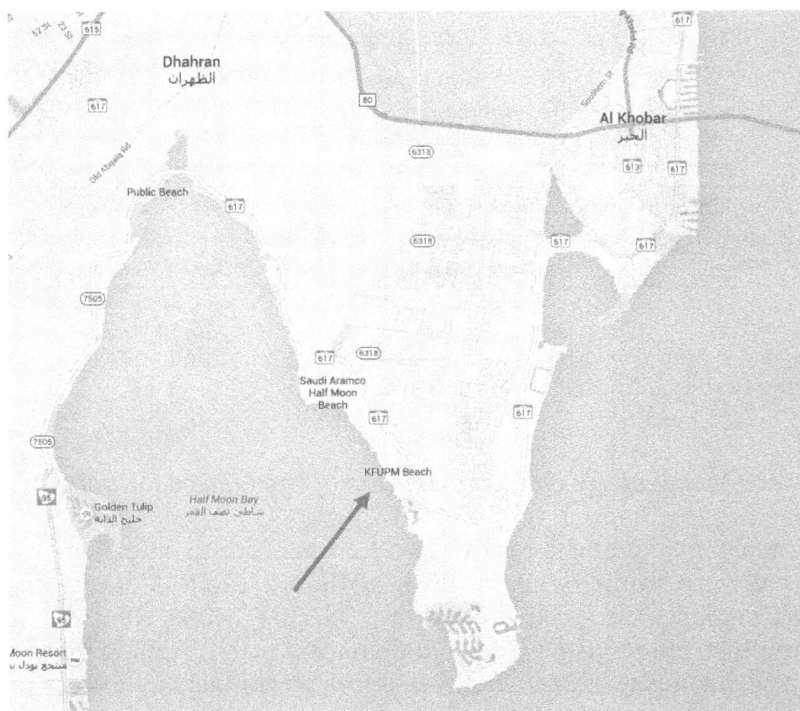

Figure 4.8 Map showing the location (as indicated by the arrow) of the field trial in Saudi Arabia (© 2018 Google Maps).

Electrochemical Impedance Spectroscopy (EIS)

A 100 ml conductivity cell (Gamry, US) was used in this study with SCE reference and graphite counter electrodes, respectively, using 3.5 wt% NaCl solution. The electrode terminals were connected to a corrosion work station (GAMRY 3000, Gamry Instruments, US). Impedance tests (EIS) were conducted on the coated panels at the first immersion hour and subsequently performed daily at corrosion potential (E_{corr}) ranging from 10 kHz to 10 mHz, with a 10 mV potential perturbation in 3.5 wt% NaCl solution. Electrochemical analyses were followed by data fitting and simulation using EChem Analyst software.

Tafel Polarization Measurements

The protective behavior against corrosion was studied also by potentiodynamic polarization in 3.5 wt% NaCl, with a testing area of 10 cm^2 (surface masks from Gamry were used to mask the analyzed surface), without stirring. The tests were controlled by a potentiostatic–galvanostatic GAMRY3000 corrosion measurement system, using a potential sweep rate of 1 mV/s and a potential range of -250 to 250 mV. Before starting the measurement, the potential of open circuit was measured until no changes were registered. The electrochemical cell used for the current study consists of a prepared coated electrode as the working electrode, a graphite rod as the counter electrode and a SCE as the reference electrode.

4.2.2 Results

Coating Characterization

The SEM images of the surfaces of the immersed inhibitor-free coated samples in 3.5 wt% NaCl medium for 10 days (Figure 4.9) showed a regular nonporous coating with no surface defects or mechanical damage. However, the SEM images of the inhibitor-loaded samples, SG-IZ and SG-IZB6, clearly showed the presence of multiple cracks (as indicated by arrows) in the coatings. This indicates that the loading of the coatings with ZAPP inhibitor had a negative impact on the homogeneity of the coating and coating performance. The SEM micrographs of cross-sections of the immersed-coated samples (Figure 4.10) revealed excellent bonding (adherence) of the applied coatings on steel surface. The thickness of the coatings obtained using the cross-sectional SEM micrographs were very close and ranged between 131 to 136 μm (Figure 4.10).

Contact angle is an important parameter for evaluating the degree of surface wetness of protective hybrid coatings [46]. Contact angle

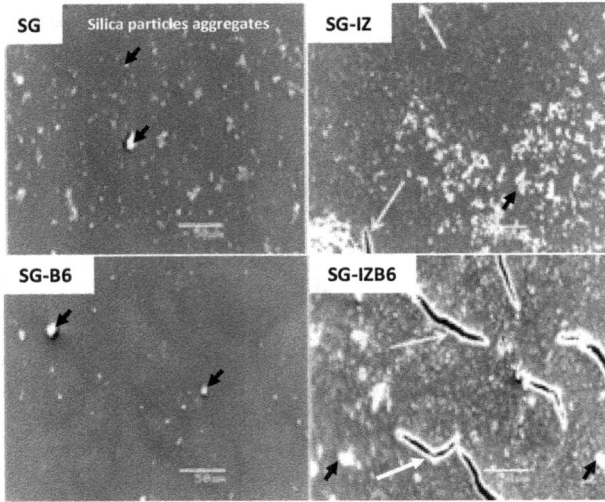

Figure 4.9 SEM micrographs of exposed sol-gel coating surfaces after a week immersion in 3.5 wt% NaCl.

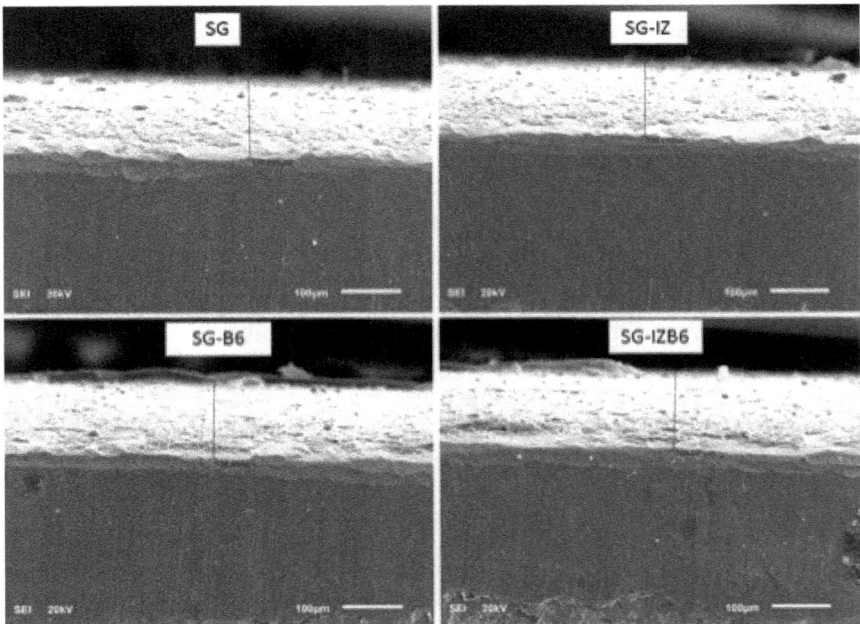

Figure 4.10 SEM cross-section micrographs of the hybrid sol-gel-coated samples after immersion in 3.5 wt% NaCl for one week.

measurements of all coating matrices before (pre) and after (post) immersion in 3.5 wt% NaCl are presented in Figure 4.11. Before exposure to the saline medium, the contact angle for the coated samples was less than 90° (Figure 4.3), revealing that the coatings were relatively hydrophilic. Moreover, the presence of endospores yielded an increase in the hydrophobicity of coatings, with the sol-gel coating containing bacterial endospores (73°) and ZAPP (70°) exhibiting the most hydrophobic surface (90°). The increase in contact angle of the coatings in the presence of the bacterial endospores could be attributed to the secretion of antibiotic type substance within their biofilms [47,48]. After the immersion testing, a remarkable decrease in the contact angle of all coatings was observed, indicating a drop in hydrophobicity, except for SG-B6 sample which showed the least wetting behavior among all samples. The loss in the hydrophobicity of the surfaces refers to an onset of corrosion behavior of the coated samples [49,50].

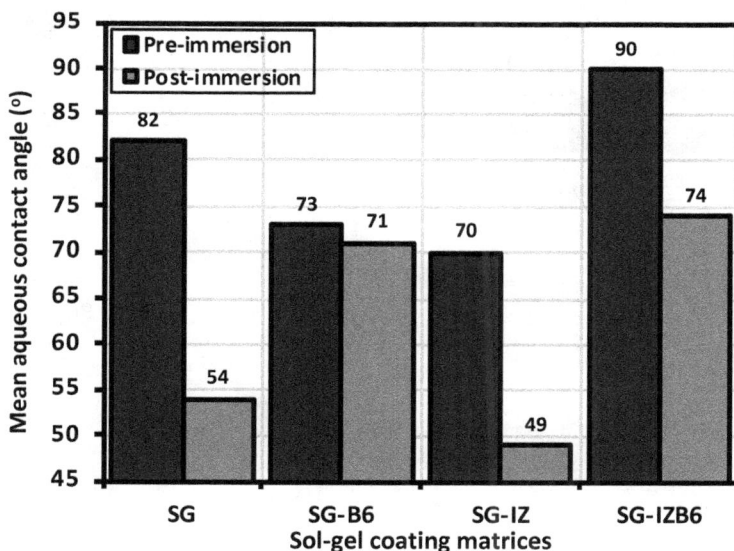

Figure 4.11 Mean contact angle of each sol-gel matrix before (pre; first column) and after (post; second column) immersion in 3.5 wt% NaCl.

Confocal Fluorescence Microscopy

Abiotic and biotic sol-gel without inhibitors and in combination with the corrosion inhibitor ZAPP were coated on glass slides, which were examined by CLSM, after staining with the BacLight Live/Dead stain (Invitrogen) after incubation in liquid nutrient medium (NB2) at 30 °C for one/two days, to observe the embedded bacteria (Figure 4.12).

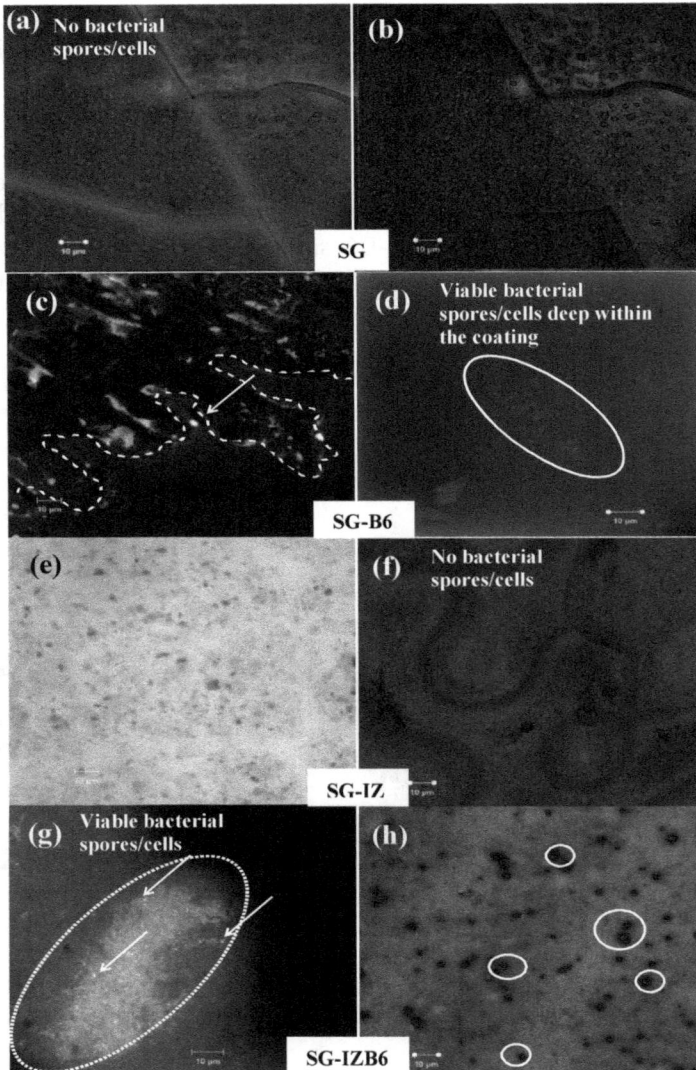

Figure 4.12 Multichannel CLSM images of the abiotic [SG: (a) & (b); SG-IZ: (e) & (f)] and biotic [SG-B6: (c) & (d); SG-IZB6: (g) & (h)] sol-gel matrices after immersion in NB2 for one (left panel) and two days (right panel).

The bacteria in the coated samples were fairly obvious as the visible endospores, as indicated by arrows in Figure 4.12g; group and individual spores are also circled in Figure 4.12d and 4.12h respectively. The sol-gel coating without inhibitor (Figure 4.12c, 24 h immersion) fragmented and was observed to partially fall off the glass substrate. The ZAPP-containing SG-IZ (Figure 4.5e and f) had a more homogeneous appearance, with an increased amount of autofluorescence resulting from the Syto 9 green stain in the sol-gel matrix in Figure 4.12e. The red propidium iodide inclusions might result from the adherence of this stain to negatively charged regions in the sol-gel, possibly within pores. In the image of the SG-IZB6 coating (Figure 4.12g, 24 h immersion), a clearly defined area containing many bright green ovals might be a cluster of germinating endospores (circled, with brighter spots arrowed). After 48 h immersion in NB2, the image of the SG-coated glass slides showed the presence of dark areas (Figure 4.12f), which appeared to contain a bright dot (representatives are circled) and could indicate germinating endospores or cells. The dark areas might represent pores in the coating.

Field Trial Results

The marine biofouling testing of the sol-gel coatings on mild steel was carried out in seawater at the Half Moon Bay (KFUPM beach), Saudi Arabia, in comparison with bare mild steel sample. The Eastern Province of Saudi Arabia is known for minimal seawater level, with high salinity and marine macro-fouling characteristics. The average salinity value measured for KFUPM beach was 68.5 ppt for the period of study between March and June 2014; this body of water could be classified as being "hyperhaline". Photographs were taken for all samples periodically for 6 weeks (Figure 4.13). The bare steel panel was instantly corroded after immersing it in the seawater, while no signs of corrosion and fouling products on the surface of coated samples were observed after 2 weeks of immersion. Various factors such as the nutrient availability, surface chemistry, topology and rheology/hydrodynamics can generally affect the initial settlement of biofouling organisms on any substrate. After 6 weeks of immersion, the inhibitor-loaded coatings failed to protect the steel surface against corrosion and fouling, as compared to inhibitor-free coatings. Moreover, the biotic SG-B6 sample showed better anticorrosion and antifouling performance, as compared to the abiotic coated sample. The field trial results also showed the assemblage of annelid-type polychaetes and

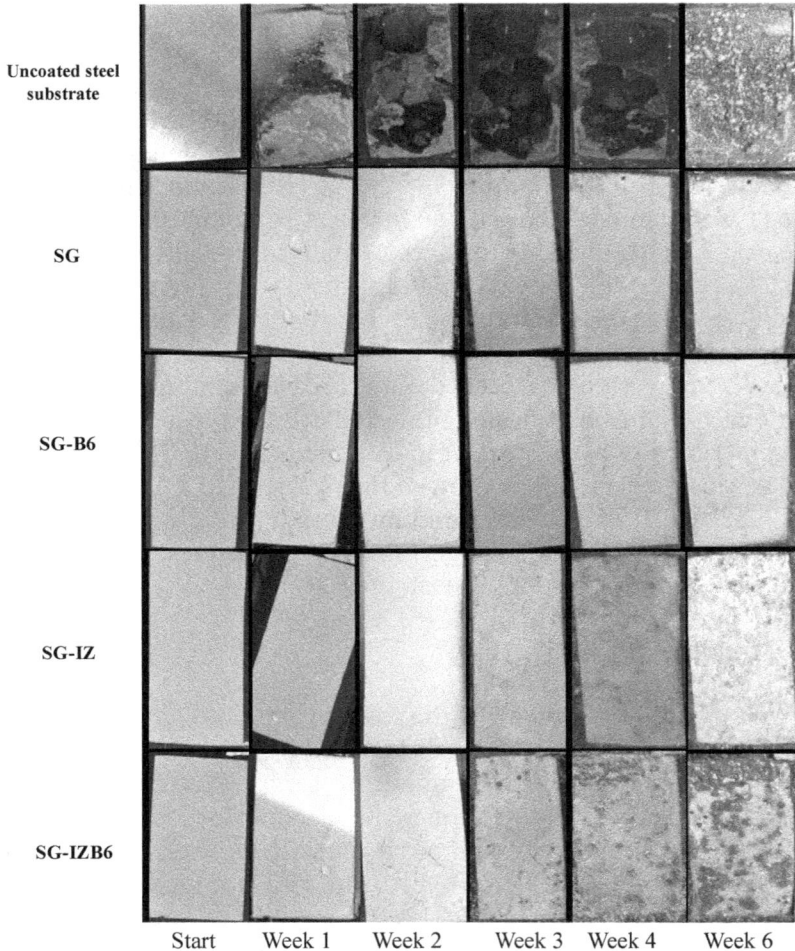

Figure 4.13 Images of field trial samples for coating matrices and bare/ uncoated steel samples immersed at Half Moon Bay, Saudi Arabia.

corrosion products on the surface of SG-IZ and SG-IZB6 sample which revealed the incompatibility of this class of sol-gel coating with the ZAPP inhibitor. The appearance of SG-IZB6 at the 6th week suggested a lack of synergy between the inhibitor and bacteria, when combined in the coating matrix. The overall analysis of the field testing results indicated that the protective sol-gel coatings immobilized with *B. licheniformis* (B6) bacterial endospores demonstrate promising antifouling activity and corrosion protection properties within the duration and conditions of study.

Electrochemical Evaluation of Protective Properties of Coated Samples

In order to assess the anticorrosion properties of the abiotic and biotic coated samples in saline medium, the steel-coated panels were immersed continuously in 3.5 wt% NaCl medium and the required electrochemical tests were conducted at specific times of immersion. The reason behind selecting this electrolytic solution in our electrochemical corrosion testing is its extensive use in literature as a source for chloride content, which is very similar to the seawater content. Figure 4.14 presents digital images for the sol-gel type coatings on mild steel after 10 days of exposure to 3.5 wt% NaCl. Signs of corrosion could be observed on the surface of SG and SG-IZ samples, and, to a lesser extent, on the surface of SG-IZB6 sample. These results also reflected in the field testing of coated samples; SG-B6 coated steel sample remained intact after long term immersion in NaCl solution, indicating an outstanding ability for this coating to prohibit the passage of corrosive ions to and from the solution to the steel surface.

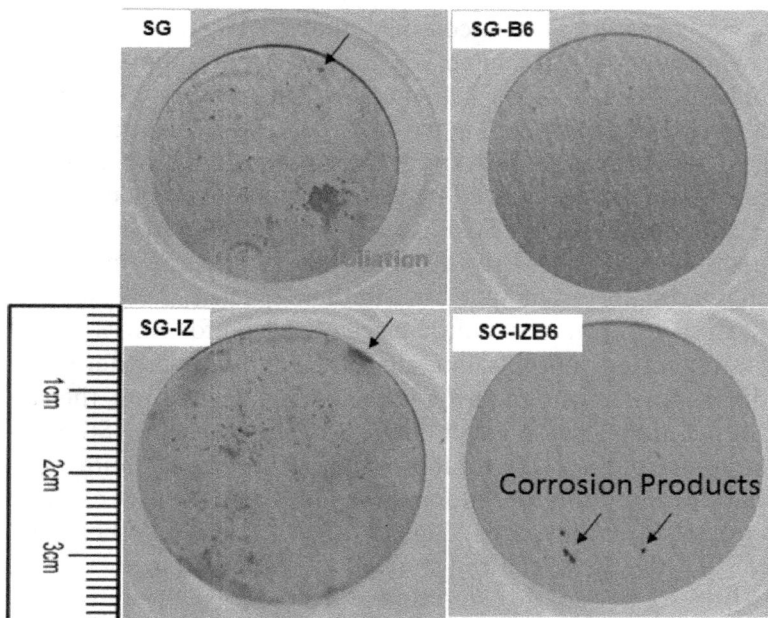

Figure 4.14 Appearance of each sol-gel coating matrix after 10 days of exposure to 3.5 wt% NaCl; samples are masked (using electrochemical mask by Gamry (US)) to give a defined exposed test area of 10 cm².

EIS Results

EIS measurements is an important technique for distinguishing the processes at the interfaces of metal/oxide layer/electrolyte in the electrochemical systems. This kind of measurement also provides a valuable quantitative assessment and characterization of the evolution processes during the exposure time [51]. The EIS data for the SG-coated steel samples immersed continuously in saline medium were collected at various intervals. This data is presented as Bode (Figure 4.15) and Nyquist (Figure 4.16) plots, respectively. Bode diagram shows the variation of impedance as a function of frequency of the electrochemical double layer formed at the sample surface in contact with the electrolyte solution. Figure 4.15a depicts the EIS Bode spectra of SG-coated samples after 24 h continuous immersion in 3.5% NaCl solution. At this immersion time, the SG-B6 sample showed the highest impedance module (1.7×10^6 Ω cm^2), while sample SG-IZ exhibited the lowest impedance properties (3.2×10^4 Ω cm^2). At longer immersion time (see Figure 4.15b), the highest impedance module corresponded to biotic samples, confirming the superior anticorrosive behavior over the abiotic coating. Figure 4.16 summarizes the Nyquist spectra of sol-gel coated samples after continuous immersion for 24 and 240 h in 3.5% NaCl solution. The Nyquist plot shows that after immersion for 24 h (Figure 4.16a), the SG-B6 coating exhibited largest semicircles, indicating greater corrosion protection. At this stage, any advantage offered by the inclusion of corrosion inhibitors was not apparent, since SG-IZ and SG-IZB6 exhibited less corrosion protection efficiencies than inhibitor free samples SG and SG-B6 (which showed significant similarities in their spectra). Two semicircles, indicating two time constants, were observed for sample SG at all immersion periods in the saline medium. Following further exposure of the samples for up to 240 h in the corrosive medium, the sol-gel coatings could be ranked as follows: SG-IZB6 > SG-B6 > SG > SG-IZ. The influence of the ZAPP inhibitor appeared to be advantageous to the coating containing B6 endospores and to be disadvantage in the absence of B6 endospores (Figure 4.16b). It can be, therefore, concluded that the SG-IZ coating showed lower impedance than the inhibitor-free sol-gel coating due to the decrease in the interfacial barrier properties in the presence of ZAPP inhibitor [36,52]. Moreover, significant deterioration on the surface of SG and SG-IZ at prolonged immersion times might have negatively affected the correlation between the EIS results and the visual appearance.

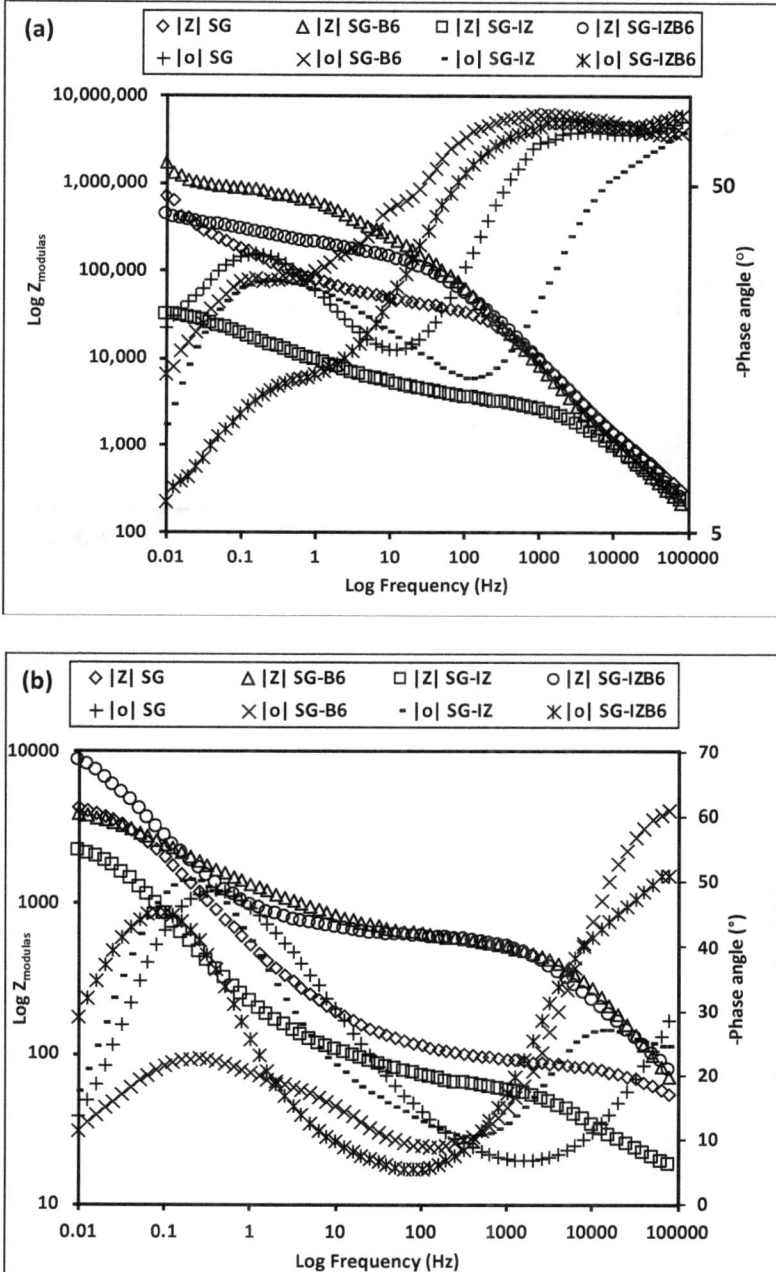

Figure 4.15 Bode plots of the sol-gel coated mild steel samples after (a) 24 and (b) 240 hours immersion in 3.5 wt% NaCl solution.

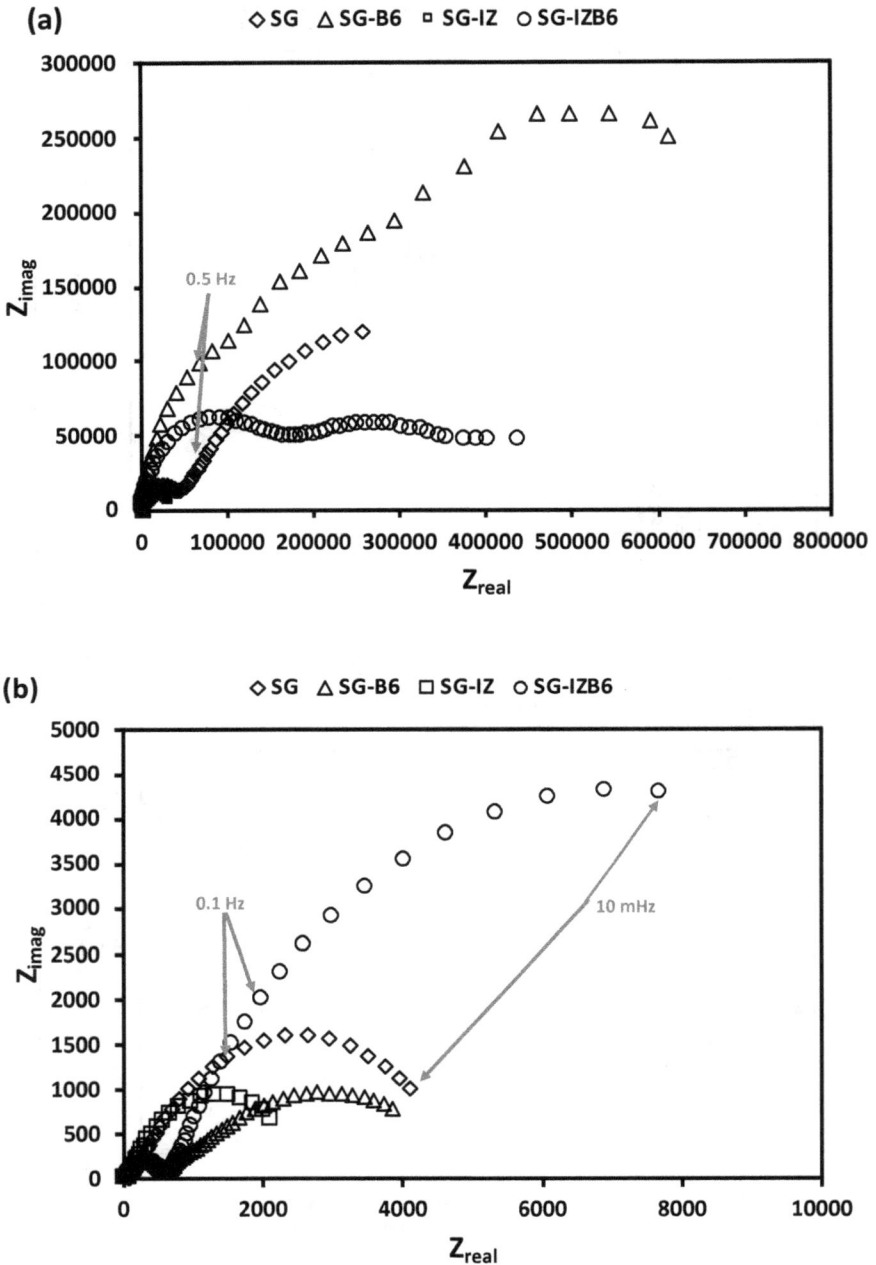

Figure 4.16 Nyquist plots of the sol-gel coated mild steel samples after (a) 24 and (b) 240 hours of exposure to 3.5 wt% NaCl solution.

Equivalent Circuit Models Used in Fitting the Experimental Data

Impedance modelling was performed in order to quantitatively inter-
pret the data obtained from the EIS experiments. Nyquist diagrams
were analyzed to determine equivalent circuits (EC) using Echem Analyst software. The equivalent circuits (Figure 4.17) are chosen based
on the number of time-constants and the quality of fit. In all cases,
apart from the SG sample, a 3-time constants hierarchically distributed equivalent circuit [denoted R(QR(QR(QR)))] was used to fit the

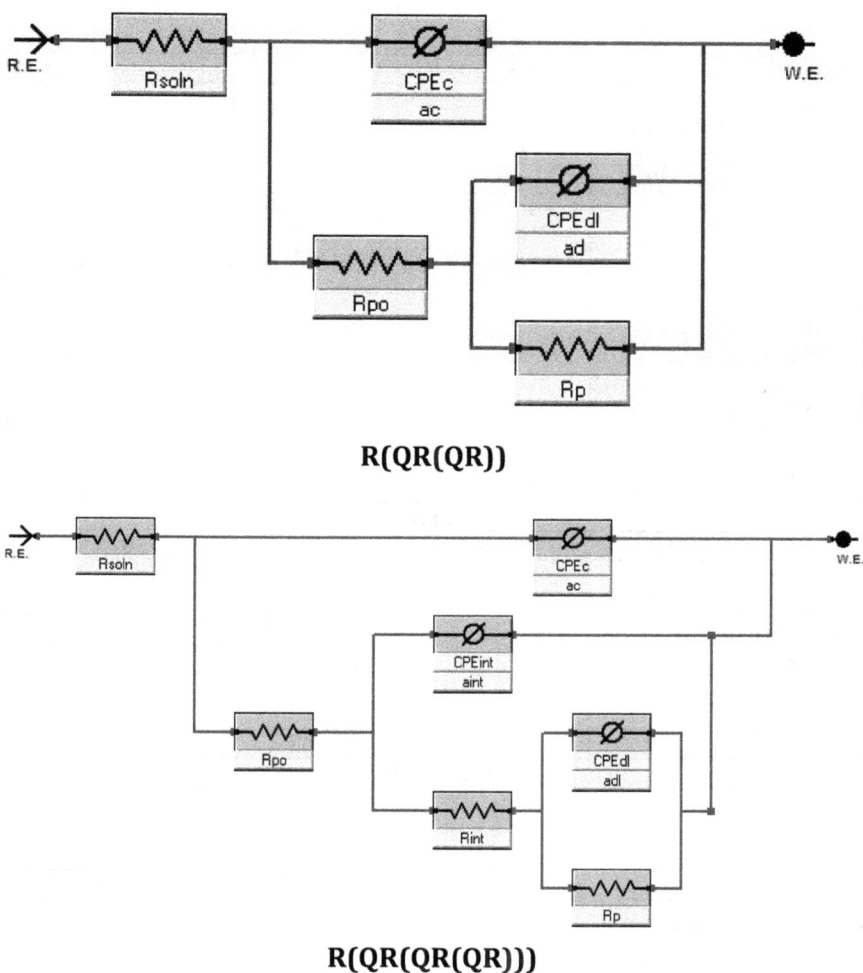

R(QR(QR))

R(QR(QR(QR)))

Figure 4.17 Equivalent circuit models used to fit experimental data.

EIS data at all immersion periods. For the SG sample, a standard paint circuit with 2-time constants [denoted R(QR(QR))] was used to fit the EIS data at all immersion periods. Parameters obtained from EIS-data fitting for all samples at various immersion times in 3.5 wt% NaCl solution are shown in Table 4.1.

Table 4.1 SG circuit fitting data

1 h				
Element	SG	SG-B6	SG-IZ	SG-IZB6
Rsoln (Ω)	10.0	5.0	10.0	5.0
Rpo (Ω)	6.4×10^3	1.6×10^3	0.2	5.2×10^5
CPEc (Fcm^{-2} s$^{-(1-\alpha c)}$)	2.6×10^{-8}	1.8×10^{-8}	2.4×10^{-8}	9.2×10^{-8}
ac	0.9	0.9	0.9	0.8
Rint (Ω)	-	1.1×10^7	2.0×10^4	3.2×10^5
CPEint (Fcm^{-2} s$^{-(1-\alpha c)}$)	-	1.6×10^{-8}	1.8×10^{-6}	9.2×10^{-7}
aint	-	0.9	0.4	0.7
Rp (Ω)	6.7×10^6	3.5×10^7	3.8×10^5	4.3×10^5
CPEdl (Fcm^{-2} s$^{-(1-\alpha c)}$)	3.7×10^{-9}	2.2×10^{-7}	1.2×10^{-5}	3.0×10^{-6}
ad	1.0	0.6	0.4	1.0
Circuit	R(QR(QR))	R(QR(QR(QR)))	R(QR(QR(QR)))	R(QR(QR(QR)))
24 h				
Element	SG	SG-B6	SG-IZ	SG-IZB6
Rsoln (Ω)	10.0	5.0	10.0	5.0
Rpo (Ω)	4.4×10^4	783.5	2.9×10^3	1.4×10^5
CPEc (Fcm^{-2} s$^{-(1-\alpha c)}$)	7.0×10^{-8}	1.9×10^{-8}	1.9×10^{-7}	6.3×10^{-8}
ac	0.8	0.9	0.8	0.8
Rint (Ω)	-	2.4×10^5	3.1×10^4	2.6×10^4
CPEint (Fcm^{-2} s$^{-(1-\alpha c)}$)	-	2.3×10^{-8}	4.9×10^{-5}	3.3×10^{-7}
aint	-	0.9	0.5	0.9
Rp (Ω)	5.0×10^5	5.9×10^5	1.2×10^4	2.4×10^5
CPEdl (Fcm^{-2} s$^{-(1-\alpha c)}$)	7.1×10^{-6}	3.1×10^{-7}	4.2×10^{-5}	5.0×10^{-6}
ad	0.6	0.7	0.9	0.7

Circuit	R(QR(QR))	R(QR(QR(QR)))	R(QR(QR(QR)))	R(QR(QR(QR)))
168 h				
Ele- ment	**SG**	**SG-B6**	**SG-IZ**	**SG-IZB6**
Rsoln (Ω)	10.0	5.0	10.0	10.0
Rpo (Ω)	383.1	5.3×10^3	305.7	1.9×10^3
CPEc (Fcm^{-2} s$^{-(1-\alpha c)}$)	9.5×10^{-7}	8.9×10^{-8}	8.6×10^{-6}	2.9×10^{-7}
ac	0.7	0.9	0.5	0.8
Rint (Ω)	-	1.5×10^4	3.2×10^3	2.3×10^3
CPEint (Fcm^{-2} s$^{-(1-\alpha c)}$)	-	3.2×10^{-5}	3.8×10^{-4}	2.0×10^{-4}
aint	-	0.5	0.6	0.6
Rp (Ω)	2.1×10^4	1.1×10^4	8.3×10^3	1.9×10^4
CPEdl (Fcm^{-2} s$^{-(1-\alpha c)}$)	1.6×10^{-4}	1.2×10^{-4}	1.6×10^{-4}	8.5×10^{-5}
ad	0.6	0.1	0.9	0.9
Circuit	R(QR(QR))	R(QR(QR(QR)))	R(QR(QR(QR)))	R(QR(QR(QR)))
240 h				
Ele- ment	**SG**	**SG-B6**	**SG-IZ**	**SG-IZB6**
Rsoln (Ω)	10.0	4.6	10.4	8.9
Rpo (Ω)	86.1	554.7	58.7	627.6
CPEc (Fcm^{-2} s$^{-(1-\alpha c)}$)	3.4×10^{-6}	4.4×10^{-7}	2.5×10^{-5}	1.8×10^{-6}
ac	0.6	0.8	0.6	0.7
Rint (Ω)	-	2.0×10^3	89.9	1.1×10^3
CPEint (Fcm^{-2} s$^{-(1-\alpha c)}$)	-	3.2×10^{-4}	4.6×10^{-4}	4.1×10^{-4}
aint	-	0.5	0.8	0.7
Rp (Ω)	2.8×10^3	2.9×10^3	2.6×10^3	1.2×10^4
CPEdl (Fcm^{-2} s$^{-(1-\alpha c)}$)	5.6×10^{-4}	4.5×10^{-4}	1.0×10^{-3}	1.4×10^{-4}
ad	0.6	0.8	0.8	0.8
Circuit	R(QR(QR))	R(QR(QR(QR)))	R(QR(QR(QR)))	R(QR(QR(QR)))

Figures 4.18 presents a representative plot of the comparison between the experimental and fitted data obtained using the equivalent circuit depicted in Figure 4.17, for sample SG-IZB6 immersed for 240 h in 3.5% NaCl solution. As can be seen from the figure, a good agreement existed between the experimental and equivalent circuit fitted data.

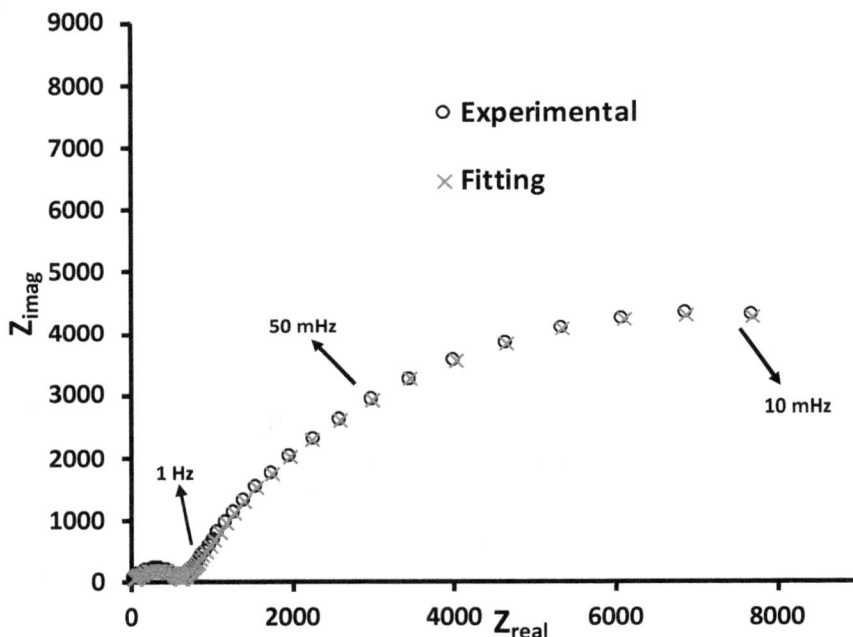

Figure 4.18 Experimental and fitted Nyquist plots for SG-IZB6 sample immersed for 240 h in 3.5% NaCl solution.

The physical interpretation of the equivalent circuits depicted in Figure 4.17 has previously been proposed [8,53-55]. The equivalent circuits comprised of a solution resistance (Rs) in series with a capacitor in the form of a constant phase element (CPEc) in parallel with a further resistor (Rpo) which represents the dielectric properties of the sol-gel layer. A constant phase element (CPE) was used instead of an "ideal" capacitor, taking into account that the slopes of the curves in the Bode plots were not -1 (a value expected for an ideal capacitor). The use of CPE as a circuit component accounts for the inhomogeneous properties of the double layers, non-uniform thickness of the coating, electrode surface irregularities (e.g. roughness and porosity) and inherent dispersive characters of time constants [56,57]. The equivalent circuits were proposed and established assuming an electrolyte resistance (Rs) and a coating layer, where the pore or defect resistance within the sol-gel layer is represented by Rpo, in parallel with CPEc, and the properties of the intermediate oxide layer are described by a resistance Rint, in parallel with a capacitor CPEint. A passive layer or film layer (represented by CPEdl and Rp) is included to represent the presence of the corrosion inhibitors or a build-up of corrosion products close to the substrate surface [58].

The fitting results reported in Table 4.1 show that the resistance decreased with exposure time for all samples, even though the values remained high at the end of the experiment. Reduced Rp values referred to the occurrence of a deterioration in the coating integrity, and the behavior of CPEc values was an indirect confirmation, by monitoring the water uptake ability of the coating. Figure 4.19 depicts the evolution of passive film capacitance with immersion time in 3.5 wt% NaCl solution for the abiotic and biotic sol-gel coated mild

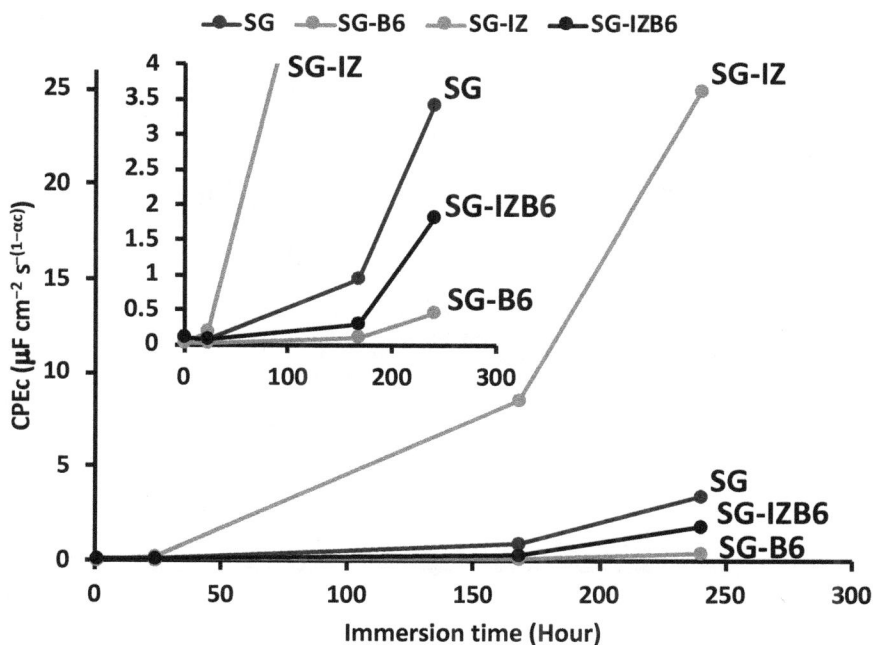

Figure 4.19 Variation in passive layer capacitances with exposure time (in hours) in 3.5 wt% NaCl solution for the sol-gel coated mild steel samples.

steel samples. An increase in capacitance with immersion time was observed, which was related to water uptake process [25]. Although this tendency was similar for all types of samples, it was stronger for the abiotic samples SG and SG-IZ, indicating a higher tendency for water uptake. Further, this tendency supported the loss of hydrophobicity trends (contact angle measurement) reported in Figure 4.11. Low values of CPEc were recorded for the SG-B6 and SG-IZB6 coatings after 240 h of continuous immersion in the saline medium denoting reduced uptake of water in the bulk of these coatings.

Potentiodynamic Polarization Results

Potentiodynamic scanning is an important tool in corrosion testing as it gives valuable details about the corrosion mechanisms and corrosion rate of metallic materials in specific medium. The current-potential curves showing the kinetics of the mild steel electrodes coated with the sol-gel coatings (after 240 h immersion in 3.5 wt% NaCl solution) for both anodic and cathodic reactions are displayed in Figure 4.20. From these polarization curves, the Tafel parameters were

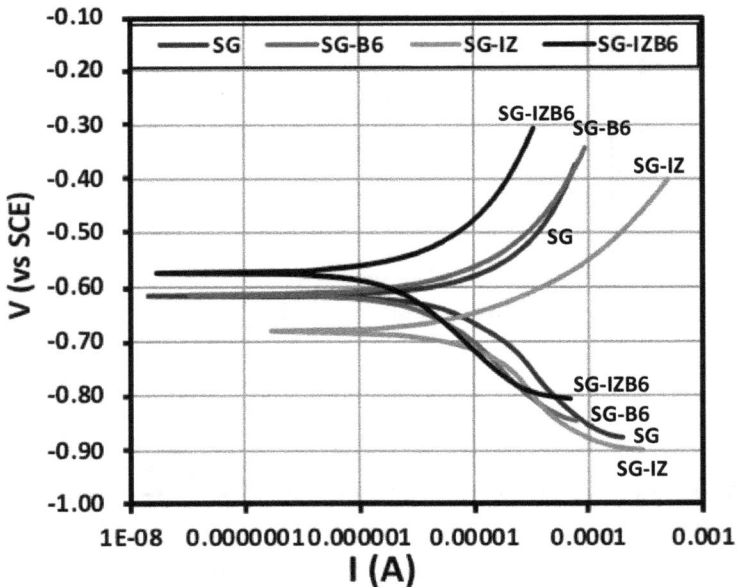

Figure 4.20 Potentiodynamic polarization curves of each biotic and abiotic sol-gel matrix coated on mild steel after 240 hours immersion in 3.5 wt% NaCl solution.

obtained and are listed in Table 4.2. The results showed that the values of corrosion current density (I_{corr}) of SG-B6 and SG-IZB6 were less than SG and SG-IZ, which confirmed the superior barrier properties of the biotic coatings over the abiotic coatings. It is also evident that the results of polarization tests were consistent with those of impedance measurements, confirming the best corrosion protection performance of SG-IZB6 and the poorest performance of SG-IZ. The magnitudes of I_{corr} as low as 2.51 and 5.18 $\mu A/cm^2$ were recorded for the sol-gel matrices modified with the bacterial endospores alone (SG-B6) and in combination with ZAPP (SG-IZB6), respectively, compared to the unmodified coating (10.4 $\mu A/cm^2$) after 240 h immersion in the solution of the electrolyte. The order of I_{corr} decrease was: SG-IZB6<SG-B6<SG<SG-IZ.

Table 4.2 Electrochemical parameters derived from the polarization curves of each biotic and abiotic sol-gel matrix coated on mild steel after 240 h immersion in 3.5 wt% NaCl solution

Coating/ parameter	SG	SG-B6	SG-IZ	SG-IZB6
E_{corr} (mV)	-614	-612	-681	-574
I_{corr} ($\mu A/cm^2$)	10.4	5.18	12.9	2.51

4.3 Conclusions

In summary, hybrid sol-gel materials have been proved to be efficient protective anticorrosion/antifouling coatings for metal substrates in the marine environment. This area of research will continue to be remarkably active in the future as the search for coatings effective against corrosion and fouling of metallic infrastructures is driven by various industrial sectors.

The field trial results of the case study reported in this chapter prove that the presence of both corrosion inhibitors and protective antifouling bacteria within the network of a sol-gel coating can significantly reduce the antifouling properties comparing to a coating containing protective bacteria alone. For instance, the coating containing *B. licheniformis* endospores without corrosion inhibitor outperforms the biotic inhibitor-containing coatings. However, in laboratory tests, the inhibitor and endospore-containing sol-gel coatings (SG-IZB6) exhibited the best corrosion protection over the test period. The low protection efficiencies of the inhibitor loaded sol-gel coatings in the marine environment can be attributed to cracks and mechanical defects observed in these coatings. The differences in the laboratory and field trial results may also be caused by the nature of the solutions to which the samples were exposed. For the field trial location at Half Moon Bay, the seawater is subject to changes in salinity, water temperature, sunshine and wave flow. Overall, within the exposure period of our field trial study, the presence of bacterial endospores within the coating resulted in improved corrosion and fouling resistance, owing to the surface hydrophobicity effects and discouragement of fouling bacteria settlement.

Acknowledgements

We thank King Fahd University of Petroleum and Minerals (KFUPM), Saudi Arabia for providing support to this project. This project has been funded by King Fahd University of Petroleum and Minerals under project no. FT161002. The authors are also thankful to Biomolecular Sciences Research Centre at Sheffield Hallam University (SHU) for providing the laser confocal images of the biotic samples.

References

1. Lim, H. L. (2012) Assessing level and effectiveness of corrosion education in the UAE. *International Journal of Corrosion*, **2012**, Article ID 785701.
2. Videla, H. A. (1989) Biological corrosion and biofilm effects on metal biodeterioration. In: *Biodeterioration Research 2*, O'Rear C. E., and Llewellyn G. C. (eds.), Springer, USA, pp. 39-50.
3. DeBaere, K., Verstraelen, H., Lemmens, L., Lenaerts, S., and Potters, G. (2011) In Situ Study of the Parameters Quantifying the Corrosion in Ballast Tanks and an Evaluation of Improving Alternatives. *NACE Conference Papers*, USA. Online: http://www.shipstructure.org/pdf/11symp02.pdf [accessed 2nd January 2019].
4. Koch, G. H., Brongers, M. P. H., Thompson, N. G., Vimani, Y. P., and Payer, J. H. (2002) *Corrosion Costs and Preventive Strategies in the United States*, US Federal Highway Administration, USA. Online: https://www.nace.org/uploadedfiles/publications/ccsupp.pdf [accessed 19th December 2018].
5. Tems, R., and Al Zahrani, A. M. (2006) Cost of corrosion in oil and refining. *Saudi Aramco Journal of Technology*, Summer Edition.
6. Sorensen, P. A., Kiil, S., Dam-Johansen, K., and Weinell, C. E. (2009) Anticorrosive coatings: a review. *Journal of Coatings Technology and*

Research, **6**, 135-176.

7. Suleiman, R., Estaitie, M., and Mizanurahman, M. (2016) Hybrid organosiloxane coatings containing epoxide precursors for protecting mild steel against corrosion in a saline medium. *Journal of Applied Polymer Science*, **133**(8), doi: 10.1002/APP.43947.

8. Eduok, U., Suleiman, R., Khaled, M., and Akid, R. (2016) Enhancing water repellency and anticorrosion properties of a hybrid silica coating on mild steel. *Progress in Organic Coatings*, **93**, 97-108.

9. Chandler, K. A. (1985) *Marine and Offshore Corrosion*, Butterworths, USA.

10. Almeida, E., Diamantino, T. C., and Sous, O. D. (2007) Marine paints: The particular case of antifouling paints, *Progress in Organic Coatings*, **59**, 2-20.

11. *Metallic Coatings*. Online: https://corrosion-doctors.org/Metal-Coatings/MetCoat.htm [accessed 21st December 2018].

12. *Corrosion Protection and Prevention*, Anochrome Group (2019). Online: https://www.anochrome.com/technical/corrosion-protection-prevention/ [accessed 19th December 2018].

13. Santana, I., Pepe, A., Schreiner, W., Pellice, S., and Cere, S. (2016) Hybrid sol-gel coatings containing clay nanoparticles for corrosion protection of mild steel. *Electrochimica Acta*, **203**, 396-403.

14. Su, H.-Y., Chen, P.-L., and Lin, C.-S. (2016) Sol-gel coatings doped with organosilane and cerium to improve the properties of hot-dip galvanized steel. *Corrosion Science*, **102**, 63-71.

15. Figueira, R. B., Silva, C. J. R., and Pereira, E. V. (2015) Hybrid sol-gel coatings for corrosion protection of hot-dip galvanized steel in alkaline medium. *Surface and Coatings Technology*, **265**, 191-204.

16. Suleiman, R., Dafalla, H., and El Ali, B. (2015) Novel hybrid epoxy silicone materials as efficient anticorrosive coatings for mild steel. *RSC Advances*, **5**, 39155-39167.

17. Daniels, M. W., and Francis, L. F. (1998) Silane adsorption behavior, microstructure, and properties of glycidoxypropyltrimethoxysilane-modified colloidal silica coatings. *Journal of Colloid and Interface Science*, **205**, 191-200.

18. Daniels, M. W., and Francis, L. F. (1999) Effect of curing strategies on porosity in silane modified silica colloidal coatings. *Materials Research Society Symposium Proceedings*, **576**, 313-317.

19. Bajpai, M., and Sharma, A. (1997) Film characteristics of novolac based epoxy ester films – II. *Paint India*, **47**, 53-58.

20. Brushwell, W. (1981) Coatings update: Epoxy resins. *Paint India*, **31**, 11-16.

21. Suleiman, R. K., Saleh, T. A., Al Hamouz, O. C. S., Ibrahim, M. B., Sorour, A. A., and El Ali, B. (2017) Corrosion and fouling protection performance of biocide-embedded hybrid organosiloxane coatings on mild steel in a saline medium. *Surface and Coatings Technology*,

324, 526-535.

22. Shakeri, A., Abdizadeh, H., and Golobostanfard, M. R. (2014) Synthesis and characterization of thick PZT films via sol-gel dip coating method. *Applied Surface Science*, **314**, 711-719.

23. Banerjee, D. A., Kessman, A. J., Cairns, D. R., and Sierros, K.A. (2014) Tribology of silicananoparticle-reinforced, hydrophobic sol-gel composite coatings. *Surface and Coatings Technology*, **260**, 214-219.

24. Eduok, U., Faye, O., and Szpunar, J. (2017) Recent developments and applications of protective silicone coatings: A review of PDMS functional materials. *Progress in Organic Coatings*, **111**, 124-163.

25. Collazo, A., Covelo, A., Novoa, X. R., and Perez, C. (2012) Corrosion protection performance of sol-gel coatings doped with red mud applied on AA2024-T3. *Progress in Organic Coatings*, **74**, 334-342.

26. Deflorian, F., Rossi, S., Fedel, M., and Motte, C. (2010) Electrochemical investigation of high-performance silane sol-gel films containing clay nanoparticles. *Progress in Organic Coatings*, **69**, 158-166.

27. Suleiman, R., Mizanurrahman, M., Alfaifi, N., El Ali, B., and Akid, R. (2013) Corrosion resistance properties of hybrid organic-inorganic epoxy-amino functionalized polysiloxane based coatings on mild steel in 3.5% NaCl solution. *Corrosion Engineering, Science and Technology*, **48**, 525-529.

28. Suleiman, R., Khaled, M., Wang, H., Smith, T. J., Gittens, J., Akid, R., El Ali, B., and Khalil, A. (2014) A comparison of selected inhibitor doped sol-gel coating systems for the protection of mild steel. *Corrosion Engineering, Science and Technology*, **49**, 189-196.

29. Galliano, F., and Landolt, D. (2002) Evaluation of corrosion protection properties of additives for waterborne epoxy coatings on steel. *Progress in Organic Coatings*, **44**, 217-225.

30. Suleiman, R., Kumar, M.A., Sorour, A., Al-Badour, F., and El Ali, B. (2018) Hybrid organosiloxane material/metal oxide composite as efficient anticorrosive coatings for mild steel in a saline medium. *Journal of Applied Polymer Science*, **135**, doi: 10.1002/app.46718.

31. Crisp, D. (1973) The role of the biologist in antifouling research. *Proceedings of the Third I.C.M.C.F.*, Northwestern University Press, USA, p. 88.

32. Kuo, P. *et al.* (1997) Interface-crashed Polishing Type of Tin-free Antifouling Coatings. *Proceedings of the Emerging Nonmetallic Materials for the Marine Environment*, USA.

33. JPCE-JPCL staff, OSHA's national emphasis program on lead: Impact on industrial coatings industry (2011). *JPCL*, October, 42.

34. Suleiman, R., Gittens, J., Khaled, M., Smith, T. J., Akid, R., El Ali, B. and Khalil, A. (2017) Assessing the anticorrosion and antifouling performances of a sol-gel coating mixed with corrosion inhibitors and immobilised bacterial endospores. *Journal for Science and Engineering*, 42, 4327-4338.

35. Akid, R., Wang, H., Smith, T. J., Greenfield, D., and Earthman, J. C. (2008) Biological functionalization of a sol-gel coating for the mitigation of microbial-induced corrosion. *Advanced Functional Materials*, **18**, 203-211.

36. Gittens, J., Wang, H., Smith, T. J., Akid, R., and Greenfield, D. (2010) Biotic Sol-gel Coating for the Inhibition of Corrosion in Seawater. *Fifth International Symposium on Advances in Corrosion Protection by Organic Coatings*, pp. 211-229.

37. Gittens, J. E., Smith, T. J., Suleiman, R., and Akid, R. (2013) Current and emerging environmentally-friendly systems for fouling control in the marine environment. *Biotechnology Advances*, **31**, 1738-1753.

38. Katz, E., and Demain, AL. (1977) The peptide antibiotics of Bacillus: chemistry, biogenesis, and possible functions. *Bacteriological reviews*, **41**, 449-474.

39. Romanenko, L. A., Uchino, M., Kalinovskaya, N. I., and Mikhailov, V. V. (2008) Isolation, phylogenetic analysis and screening of marine mollusc-associated bacteria for antimicrobial, hemolytic and surface activities. *Microbiological Research*, **163**, 633-644.

40. Gomaa, E. Z. (2013) Antimicrobial activity of a biosurfactant produced by Bacillus licheniformis strain M104 grown on whey. *Brazilian Archives of Biology and Technology*, **56**, 259-268.

41. Abu Sayem, S. M., Manzo, E., Ciavatta, L., Tramice, A., Cordone, A., Zanfardino, A., De Felice, M., and Varcamonti, M. (2011) Anti-biofilm activity of an exopolysaccharide from a sponge-associated strain of bacillus licheniformis. *Microbial Cell Factories*, **10**, 1-12.

42. Eduok, U. M., Suleiman, R. K., Gittens, J., Khaled, M., Smith, T. J., Akid, R., El Ali, B., and Khalil, A. (2015) Anticorrosion/antifouling properties of bacterial spore-loaded sol-gel type coating for mild steel in saline marine condition: A case of thermophilic strain of Bacillus licheniformis. *RSC Advances*, **5**, 93818-93830.

43. Khalil, A. (2011) Isolation and characterization of three thermophilic bacterial strains (lipase, cellulose and amylase producers) from hot springs in Saudi Arabia. *African Journal of Biotechnology*, **10**, 8834-8839.

44. Khalil, A., Anfoka, G., and Bdour, S. (2003) Isolation of plasmids present in thermophilic strains from hot springs in Jordan. *World Journal of Microbiology and Biotechnology*, **19**, 239-241.

45. Stewart, G., Johnstone, K., Hagelberg, E., and Ellar, D. J. (1981) Commitment of bacterial spores to germinate. A measure of the trigger reaction. *Biochemical Journal*, **198**, 101-106.

46. Zhang, X., Zheng, F., Ye, L., Xiong, P., Yan, L., Yang, W., and Jiang, B. (2014) A one-pot sol-gel process to prepare a superhydrophobic and environment-resistant thin film from ORMOSIL nanoparticles. *RSC Advances*, **4**, 9838-9841.

47. Ornek, D., Jayaraman, A., Syrett, B. C., Hsu, C., Mansfeld, F., Wood, T.

(2002) Pitting corrosion inhibition of aluminum 2024 by Bacillus biofilms secreting polyaspartate or γ-polyglutamate. *Applied Microbiology and Biotechnology*, **58**, 651-657.

48. Batrakov, S. G., Rodionov, T. A., Esipov, S. E., Polyakov, N. B., Sheichenko, V. I., Shekhovtsova, N. V., Lukin, S. M., Panikov, N. S., and Nikolaev, Y. A. (2003) A novel lipopeptide, an inhibitor of bacterial adhesion, from the thermophilic and halotolerant subsurface Bacillus licheniformis strain 603. *Biochimica et Biophysica Acta*, **1634**, 107-115.

49. Jeon, H. R., Park, J. H., and Shon, M. Y. (2013) Corrosion protection by epoxy coating containing multi-walled carbon nanotubes. *Journal of Industrial and Engineering Chemistry*, **19**, 849-853.

50. Suleiman, R., Khaled, M., Khalil, A., and El Ali, B. (2018) Marine Anticorrosion and Antifouling Assessment of Multifunctionalized Hybrid Sol-Gel Coatings. *CORROSION2018 Conference and Expo*, USA, paper No. C2018-10539, pp. 1-13.

51. Matter, E. A., Kozhukharov, S., Machkova, M., and Kozhukharov, V. (2013) Electrochemical studies on the corrosion inhibition of AA2024 aluminium alloy by rare earth ammonium nitrates in 3.5% NaCl solutions. *Materials and Corrosion*, **64**, 408-414.

52. Suleiman, R. (2014) Corrosion protective performance of epoxy-amino branched polydimethylsiloxane hybrid coatings on mild steel. *Anti-Corrosion Methods and Materials*, **61**, 423-430.

53. Beiro, M., Collazo, A., Izquierdo, M., Nóvoa, X. R., and Pérez, C. (2003) Characterisation of barrier properties of organic paints: the zinc phosphate effectiveness. *Progress in Organic Coatings*, **46**, 97-106.

54. Collazo, A., Nóvoa, X. R., Peez, C., and Puga, B. (2008) EIS study of the rust converter effectiveness under different conditions. *Electrochimica Acta*, **53**, 7565-7574.

55. Alvarez, D., Collazo, A., Novoa, X. R., and Perez, C. (2014) Electrochemical behavior of organic/inorganic films applied on tinplate in different aggressive media. *Progress in Organic Coatings,* **77**, 2066-2075.

56. Yasakau, K. A., Carneiro, J., Zheludkevich, M. L., and Ferreira, M. G. S. (2014) Influence of sol-gel process parameters on the protection properties of sol-gel coatings applied on AA2024. *Surface Coatings and Technology*, **246**, 6-16.

57. Kirtay, S. (2014) Preparation of hybrid silica sol-gel coatings on mild steel surfaces and evaluation of their corrosion resistance. *Progress in Organic Coatings*, **77**, 1861-1866.

58. Suay, J. J., Rodriguez, M. T., Izquierdo, R., Kudama, A. H., and Saura, J. J. (2003) Rapid assessment of automotive epoxy primers by electrochemical techniques. *Journal of Coatings Technology*, **75**, 103-111.

5

Polymers and Polymer Composite Coatings for Marine Applications: A Review

Swati Singh and Vikas Mittal*,**

Department of Chemical Engineering, The Petroleum Institute (part of Khalifa University of Science and Technology), Abu Dhabi, UAE

*Corresponding author: vik.mittal@gmail.com
**Current address: *Bletchington, Wellington County, Australia*

5.1 Introduction

A large number of research studies have focused on the development of polymer nanocomposites by integrating polymers with nanoscale fillers [1,2]. Since nanoparticles offer extensive interfacial
contact for polymer-filler interactions, this approach results in significantly enhanced properties of polymer nanocomposites [3-5]. Owing
to the better insights at nanoscale, designing and monitoring of morphology for specific applications has also become possible. Coatings
represent one of the important fields of application of nanotechnology, as novel coatings with desired characteristics can be obtained.
For instance, it is known that when a structure is nano-sized, it may
change the wetting behavior at the surface, known as Lotus effect,
thus, resulting in super hydrophobic and self-cleaning coatings [6,7].
Several techniques have been developed for using polymers such as
poly(dimethylsiloxane) (PDMS) for generating nanocomposite coatings [8-14], such as laser etching [8], nano-casting [9] or CO_2-pulsed
lasers [10], etc. In one such example, Beigbeder *et al.* [14] reported
the synthesis and characterization of coatings based on silicone,
which were integrated with low concentration of natural sepiolite or
synthetic multi-walled carbon nanotubes. The crosslinking density,
wettability of the coatings before and after immersion as well as elastic modulus were discussed. The fouling-release (FR) and anti-fouling
(AF) behavior of the coatings were also investigated by laboratory
experiments using representative hard-fouling (Balanus) and soft-

Marine Coatings and Membranes, edited by Vikas Mittal
© 2019 Central West Publishing, Australia

fouling (Ulva) organisms. Natural sepiolite with unit cell formula $Si_{12}Mg_8O_{30}(OH)_4$ $(H_2O)_4$-$8H_2O$ is a microcrystalline-hydrated magnesium silicate with a microfibrous morphology. Natural sepiolite grows structurally through tunnels and blocks that grow up in the axial direction. Usually these are congregated as stacks of agglomerated fibers. Owing to its structural design and morphology, sepiolite has gained significant research attention for strengthening polymer matrices [11-14].

Non-stick FR coatings, which include fluoropolymers and silicon compounds, offer an eco-friendly alternative that can be used for prolonged periods [15]. FR coating technology acts by preventing fouling settlements and providing extremely smooth self-cleaning surfaces [16]. Organo-silicone polymers, particularly PDMS, are more efficient than fluoropolymers and, thus, are considered the most promising FR coating systems [17-22]. As a non-stick FR layer, PDMS with Si-O backbone and CH_3 side chains imparts excellent properties, such as high extent of smoothness and hydrophobicity, along with low surface tension and porosity [22,23]. In addition, PDMS exhibits excellent heat resistance as well as anti-oxidation and anti-ozone properties, along with durability against ultraviolet (UV) irradiation [24]. Although PDMS coatings show inherently superior FR attributes, its combination with inorganic nano-additives for enhanced performance has become necessary. Several characteristics of nanomaterials can influence the nanocomposite FR features, such as filler type, size, shape, degree of dispersion and compatibility with matrix [25-27]. Reduced surface tension and increased contact angle (CA) are also major factors that diminish fouling settlements and bacterial cohesion [28]. Developing coating formulations with noble metal nanoparticles (NPs) and metal oxides is possibly the most effective method to prevent fouling because of their stability against microbial attacks [28-30]. Among these materials, Cu_2O, ZnO, TiO_2 and Ag NPs can be easily prepared from low-cost natural sources [30].

Biofouling is a complex biochemical phenomenon. For example, diatoms bind through hydrophilic proteins, while barnacles bind through hydrophobic adhesive proteins. Hence, hydrophilic (and amphiphilic) coatings are also of interest in AF coatings design because they reduce bioadhesion and weaken protein adsorption. Overall, a series of uniform *in-situ*, *ex-situ* and sol-gel FR polymer nanocomposite coatings with varying degrees of hydrophobicity and hydrophilicity have been investigated in micro- and macro-biofouling assays and field tests [18,31,32].

5.2 Marine Biofouling

Marine biofouling involves a vast diversity of fouling organisms. It is reported that more than 4000 species have been identified on fouled structures worldwide [33,34]. Fouling organisms range in size from micrometers (bacteria, diatoms, algae spores, etc.) to centimeters (barnacles, tubeworms, oysters, mussels, etc.) [35]. It is often stated that fouling community development generally follows a linear 'successional' model [34,36,37]. In the first stage, organic molecules, such as protein, polysaccharides and proteoglycans, adhere to the surface via physical forces. Next, reversible 'adsorption' of bacteria and diatoms occurs, followed by attachment of spores of microalgae and protozoans to form a biofilm or slime layer. In the last stage, larvae of macroorganisms settle on the surface in 2-3 weeks of immersion and subsequently set for growth of either macroalgae or marine invertebrates [34,36-38]. However, the sequence of biofouling may not strictly follow the linear successional model [35,39,40], as the formation of biofilm is not a prerequisite and other marine species may settle at the same time [33]. In a more balanced view, fouling ranges from successional to non-successional processes driven by the numbers and kinds of propagules in the water column [35,39,40]. The development of marine biofouling communities is driven by several factors, including temperature, nutrient levels, flow rates, salinity and pH of the marine environment, along with the properties of material surfaces [39,41,42]. Thus, apart from the environmental factors, surface properties, such as surface energy and wettability [41-48], charge [47,49-56] and topography [57-63], have been shown to affect the adhesion of marine organisms. Many studies have demonstrated that surfaces with surface energy values between 20 and 30 mJ m^{-2}, known as the 'Baier minimum', show minimal adhesion [41,43,44,64,65]. In addition to surface energy, surface charge also plays an important role in the adhesion of micro- and macro-foulers [47,49-56]. Besides surface chemistry, surfaces with complex topographies, such as those mimicking the skin of sharks and lotus leaves, exhibit biofouling inhibition or 'self-cleaning' properties [57-61,66].

As an example, a biodegradable polyurethane with N-(2,4,6-trichlorophenyl)maleimide (TCPM) pendant groups was prepared through a combination of a thiol-ene click reaction and a condensation reaction. The TCPM moieties acting as anti-foulants were observed to be released as polymer degraded in the marine conditions (Figure 5.1) [50].

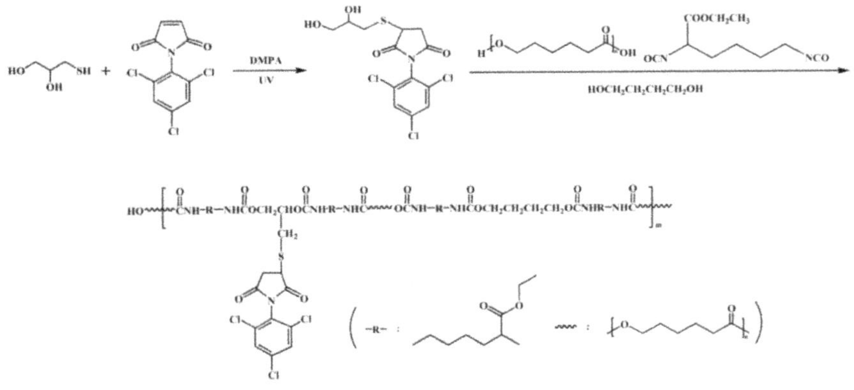

Figure 5.1 Synthesis of biodegradable polyurethane with anti-foulant pendant groups. Reproduced from Reference 50 with permission from American Chemical Society.

5.3 Polymers for Marine Applications

Biodegradable polymers are the promising materials for anti-biofouling as the erosion and degradation by enzymes or sea water can result in a surface with self-polishing behavior [67-70]. On the other hand, they can act as release and carrier systems for the anti-foulants [71-73]. Zhang *et al.* [76] reported decomposable anti-fouling systems through chemical and physical alteration of poly(ε-caprolactone) (PCL). By having a monitored release of organic anti-foulant, an excellent anti-fouling property was exhibited [74-77]. The bio-decomposition rate of the polymer is certainly important as it relates to the release rate of anti-foulants along with the rate of erosion of the surface. Another biodegradable polymer with monomer produced from renewable resources is poly(butylene succinate) (PBS) [78-80]. An elevated density of ester groups is present in the main chain of PBS with respect to PCL. The PBS and PCL homopolymers show good anti-biofouling property due to their slow decomposition owing to large spherulites and high crystallinity [77,81]. Nevertheless, the crystallinity of PCL might decrease by blending with PBS, which will result in the rapid rate of release and degradation of anti-foulants. As a matter of the fact, the rate of degradation can be regulated by the composition of the blend. Thus, in marine anti-biofouling for the governed release of anti-foulant, the role of polymer carrier is significant. In a recent study, Chen *et al.* [82] fabricated PBS and PCL blend and applied it as a carrier of organic anti-foulant (4,5-dichloro-2-octyl-isothiazolone). Differential scanning calorimetry and polarizing optical microscopy results displayed that the size of spherulite of PCL reduced and a slight change in crystallinity took place when the fraction of PBS was increased. The authors observed that amorphous interfacial regions permitted the degradation of the blend at a steady rate and the uniform dispersion of anti-foulants in the blend. The tests for marine uses confirmed good performance of the system for marine anti-biofouling [82]. In another study, Movahedi *et al.* [83] demonstrated the redox properties and absorption capacity of microparticles of poly(methylmethacrylate) (PMMA) and poly (tris[(benzyltriazol)methyl]amine) (poly(TBTA)) in the presence of copper ions. The microparticles of poly(TBTA) could absorb almost 5 wt% of copper, which corresponded to approximately 50-70% of the theoretical capability. The hybrid PMMA:poly(TBTA) microparticles also displayed copper-coordinating behavior in the coating, which was maintained after integrating inside a coating matrix like PMMA.

5.4 Polymer Nanocomposites based Fouling-resistant Coatings for Marine Applications

Amphiphilic polymer nanocomposites result in effective fouling-resistant polymer coatings. These materials possess low polymer-water interfacial energies. The high hydration degree increases the energetic penalty of removing water when bio-foulants attach to a surface. As a result, the surface becomes resistant to protein adsorption and to settlement of fouling organisms [84]. Several hydrophilic polymer nanocomposite coatings based on polyethylene glycol (PEG), hydrogels, zwitterions and hyperbranched polymers have been developed as marine AF coatings.

5.4.1 Hydrogel based Nanocomposite Coatings

Hydrogels are composed of hydrophilic polymer networks and are distinguished from solid materials by their high water composition. These hydrogels are porous three-dimensional network structures that contain 80% water. Also, they are non-toxic, highly elastic and inert against bio-macromolecular adhesion that may resist the irreversible protein fouling [85,86]. These key factors inhibit coupling formation between the cement protein and hydrogel surface, thus, prohibiting fouling attachment. Hydrogels composed of long PEG chains outperform coatings with short PEG chains because of their enhanced capability to prevent the attachment of fouling organisms. Thiol-ene click reaction has been facilitated to prepare PEG hydrogel coatings with different structural compositions, chain lengths, vinyl end groups and thiol cross-linkers [86]. Although hydrogels provide several advantages, such as efficient mass transfer, hydrophilicity and stimulus- and cell-induced responsiveness, their widespread use is hampered because of their poor mechanical properties and brittleness upon dehydration [87]. Different strategies have been used to overcome these drawbacks. For instance, hydrogel nanocomposites, such as nanoclay in hydrogel polymers, is a promising approach to generate mechanically robust hydrogels [86]. Hydrogel nanocomposite of carboxybetaine methacrylamide and 2-hydroxyethyl methacrylate with clay as reinforcement were observed to exhibit high AF and mechanical characteristics [88]. Furthermore, nanocomposite hydrogels with interpenetrating polymer network structures based on PEG methyl ether methacrylate (PEGMA)-modified ZnO NPs and 4-azidobenzoic agarose exhibited excellent mechanical and AF performance with negligible cytotoxicity [88]. These hydrogels manifest high resistance to protein adsorption, cell cohesion and bacterial settlement, thus, presenting high non-fouling behavior.

5.4.2 PEG based Fouling-resistant Nanocomposite Coatings

PEGylated materials have been applied because of their strong AF tendency against cell and protein cohesion. PEG is non-toxic, highly hydrophilic and neutrally charged material. PEG presents weak basic ether linkage and reduced interfacial energy with water (5 mJ m^{-2}), and these characteristics facilitate its good AF performance [89]. Maximizing the surface hydrophilicity and minimizing the attraction forces (caused by formation of hydrogen bonds with water) with fouling community are the mechanistic key issues of PEG [89,90]. PEG coatings suffer from rapid auto-oxidation in the presence of oxygen and transition metal ions, thereby leading to the decomposition of coatings. However, the formation of PEG nanocomposites significantly overcomes this drawback [91]. For example, PEG-ZnO nanocomposites reduced the protein adsorption by 30%. This FR nanocoating was more effective than the silicone hydrogel coating with Ag-polyvinylpyrrolidone. Furthermore, most bacteria attached on ZnO-PEG nano-surface could be eliminated after 4 h of incubation period, whereas those on Ag-polyvinylpyrrolidone based coating were effectively eliminated after 8 h [91]. PEG-grafted multiwall carbon nanotubes (MWCNTs) nanocomposites have also been synthesized and used as an AF agent to prepare nanohybrid polyethersulfone (PSf) surfaces. PEG-g-MWCNTs/PSf nanohybrid systems with effective hydrophilicity and AF performances could be applied to water purification technologies [90].

5.4.3 Hyperbranched Polymers based Anti-fouling Nanocomposite Coatings

Highly branched coating matrixes with hydrophilic branches have been extensively investigated to provide biofouling resistance to surfaces. Hyperbranched polymers are advantageous for FR applications because of numerous branches, variable branching density, high solubility, low viscosity and low volatile organic compounds (VOC) [92]. These polymers can form extremely hydrophilic surfaces resembling a hydrogel surface during water contact. Other advantages of these polymers include simple preparation through single-pot technique and purification, thus, the resulting polymer is a low-cost material [93]. Hyperbranched polyethyleneimine with tuned surface topology shows excellent AF performance and high resistance against non-specific protein adsorption [93c]. Consequently, hyperbranched polymers based nanocomposites provide surfaces with compositional and topographical complexities that hamper any favorable interactions with adhesive biomacromolecular segments secreted by marine organisms. For instance, sericin is a natural hydrophilic protein and contains polar –OH, C=O and –NH– side chains [91]. AF tests have demonstrated that sericin thin-film composite shows a more efficient fouling resistance than commercial composite surfaces. This superior performance inevitably results in an increased electrostatic repellency by sericin based thin-film composite towards fouling organisms [94].

Deka *et al.* [94b] demonstrated that hyperbranched polyurethane incorporated with Ag exhibited higher AF properties than linear analogs. Ag particles caused cellular damage through production of reactive oxygen species. Nanocomposites of catecholic hyperbranched polyglycerol with TiO_2 NPs were also observed to exhibit significant AF performance [94c]. By applying this design, the protein adsorption deceased with an increase in catechol functionality, which was also supported by the uniform dispersion of TiO_2 NPs in the polymer matrix [95].

5.4.4 Polyzwitterionic Polymers based Anti-fouling Nanocomposite Coatings

Zwitterionic polymers have been widely explored as a new generation of fouling-resistant materials. These polymers comprise positive and negative charges, which produce more potent and stabilized ionic bonds with water molecules than those created with other hydrophilic materials [84]. Thus, the zwitterionic polymers are promising fouling-resistant materials because of their excellent hydration capacity with strong hydrophilicity. Advanced easy-cleaning system has also been designed successfully by tethering a zwitterionic matrix poly(4-(2-sulfoethyl)-1-(4-vinylbenzyl)   pyridinium   betaine)

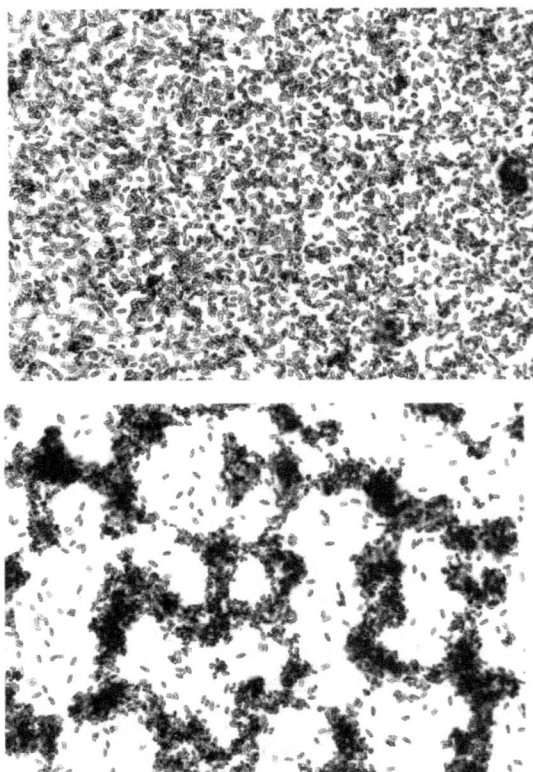

Figure 5.2 Microscopic images of cells of the diatom Navicula cultured on glass (top) and polySBMA (below) surfaces after 8 days of growth. Reproduced from Reference 97a with permission from American Chemical Society.

(PSVBP) onto polyamide (PA) surfaces [96,97]. PA-grafted-PSVBP exhibited prominent short-lived fouling prevention and salt-responsive property. The surfaces could regain self-cleaning abilities by rinsing them with brine. Zhang *et al.* [97a] also reported poly(sulfobetaine methacrylate) (polySBMA) brushes grafted onto glass surfaces using surface-initiated atom transfer radical polymerization (ATRP). Figure 5.2 also shows microscopic images of cells of the diatom Navicula cultured on glass and polySBMA surfaces.

Zwitterionic nanocomposites with controlled NP dispersion and absence of agglomerations also enhance the fouling resistance of hydrophobic materials. For instance, zwitterionic polymer brushes were developed for attachment to indium tin oxide (ITO) substrate [97c]. Photochemically grafted hydroxyl-terminated organic layers exhibited excellent AF properties (Figure 5.3). Exfoliated montmorillonite nanocomposites with catechol/zwitterionic quaternized polymer also presented AF properties and resistance against physical damage [97d]. Furthermore, the modification of zwitterionic coating films with SiO_2 NPs can significantly enhance the AF performance [98]. Eco-friendly silver-zwitterion nanocomposites were reported by Hu *et al.* [98b], which exhibited considerable antimicrobial activity and anti-adhesion characteristics (Figure 5.4).

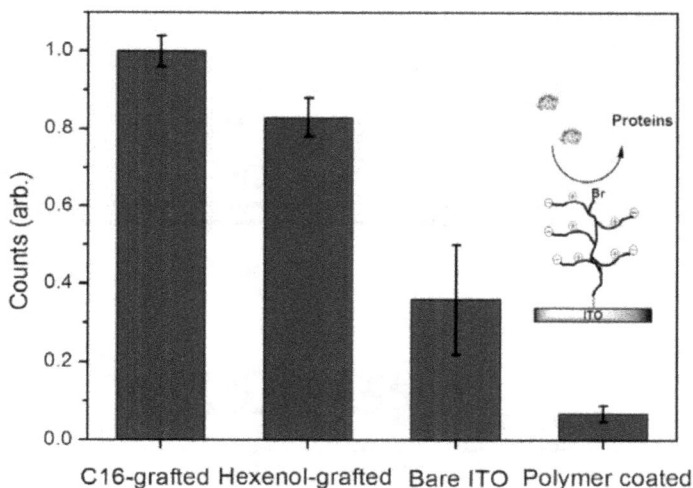

Figure 5.3 Spectroscopic evaluation of the adsorption of fluorescein isothiocyanate-labeled bovine serum albumin on different modified ITO surfaces. Reproduced from Reference 97c with permission from American Chemical Society.

Marine Coatings and Membranes

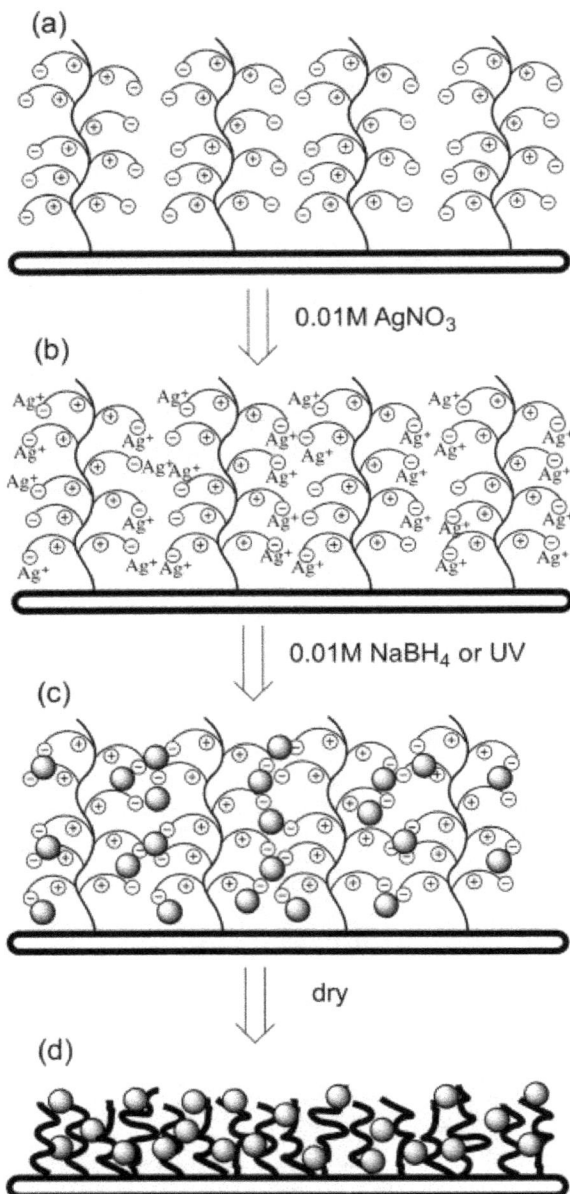

Figure 5.4 (a-c) Schematic outline of the procedure used to synthesize an Ag nanoparticle inside a zwitterionic polymer brush, (d) schematic illustration of the morphology of the silver-nanoparticle-loaded surface under dry conditions. Reproduced from Reference 98b with permission from American Chemical Society.

5.5 Polymer Brush Coatings for Combating Marine Biofouling

In a representative study, Wang *et al.* [99] reported the synthesis of polymer brushes of PEGMA and zwitterionic monomer 2-methacryloyloxyethyl phosphorylcholine (MPC) for marine antifouling. Using different fractions of crosslinker inside the feed, the functional brushes were synthesized through surface-initiated atom transfer radical polymerization (SI-ATRP). The surface became smoother and the thickness of the grafted layer enhanced by increasing the density of crosslinking. As compared to PMPC-grafted surfaces, PPEGMA-grafted surfaces showed a higher efficacy for anti-fouling.

5.6 Other Examples

In marine antifouling, it is important to have a controlled release of the anti-foulant. Chen *et al.* [100] synthesized polyurethane having decomposable polyester segments comprising poly(1,6-hexamethylene adipate) (PHA), poly(1,4-butylene adipate) (PBA) or poly(ethylene adipate) (PEA) and utilized it as release system for anti-foulants. With an increase in the decomposable segments, the rate of decomposition of polyurethane increased. However, as the segments altered from PEA, PBA to PHA, the ester group density reduced along with a decrease in the degradation rate as crystallinity escalated. It was revealed by the marine field examinations that the anti-foulant system based on polyurethane had good anti-fouling behavior. Especially, an excellent adhesion of decomposable polyurethane was observed towards the substrate which ensured longer time span of the system [100]. Due to their advantageous mechanical behaviors and anti-salt properties, basalt fiber-reinforced polymer (BFRP) tendons are also a promising prestressing component under marine environments [101].

An attempt was made to develop zwitterionic triblock copolymer (ZTC, poly(SBMA)-block-PDMS-block-poly(SBMA)) – polyurethane (PU) (ZTC-PU) coatings for marine applications [102]. The coatings were observed to be thermally stable and hydrophobic in nature. Further, the coatings showed significant non-specific marine bacterial inhibition properties as well as good stability against gram positive (S. aureus) and gram negative (E. coli and P. aeruginosa) microbial strains. In other studies, attention has also been focused on the monitoring of the initial phase of biofilm development, such as settling of benthic diatoms, bacteria and fungi [103-105]. Naamani *et al.* [106]

studied the feasibility of utilizing chitosan and ZnO oxide based hybrid coatings for fouling prevention. From the compositional study and surface morphology of the coatings, a successful interaction between the ZnO nanoparticles and chitosan was observed. In a recent study, Yee *et al.* [107] also reported a new approach to prepare nanocomposites based on graphene and silver for use as potential marine anti-fouling agent. Sathya *et al.* [108] developed hydrophobic PMMA films with a contact angle of approximately 108° using a simple ambient temperature drop-cast technique. Additionally, the nanocomposites of PMMA were developed by incorporating zinc oxide, copper oxide and cetyltrimethyl ammonium bromide capped copper oxide. Nanocomposites successfully obstructed (100%) metamorphosis and barnacle settlement in exposed cypris larvae and exhibited mortality to the level of 22 to 44%.

5.7 Conclusions

Polymer composites represent a useful class of materials for the development of AF and FR coatings for various substrates. The large variety of polymers and nanofillers makes it possible to generate numerous functional coating systems with optimal performance. Such coatings have been observed to be effective in harsh marine environments for longer periods of time.

References

1. Alexandre, M., and Dubois, P. (2000) Polymer-layered silicate nanocomposites: preparation, properties and uses of a new class of materials. *Materials Science and Engineering: R: Reports*, **28**, 1-63.
2. Schmidt, D., Shah, D., and Giannelis, E. P. (2000) New advances in polymer/layered silicates nanocomposites. *Current Opinion in Solid State & Materials Science*, **6**, 205-212.
3. De Paiva, L. B., Morales, A. R., and Guimaroes, T. R. (2007) Structural and optical properties of polypropylene-montmorillonite nanocomposites. *Materials Science and Engineering: A*, **447**, 261-265.
4. Beyer, G. (2002) Nanocomposites: a new class of flame retardants for polymers. *Plastics, Additives and Compounding*, **4**, 22-28.
5. Chaterjee, S., Goyal, A., and Shah, S. I. (2006) Inorganic nanocomposites for the next generation photovoltaics. *Materials Letters*, **60**, 3541-3543.
6. Genzer, J., and Efimenko, K. (2006) Review: recent developments in

superhydrophobic surfaces and their relevance to marine fouling. *Biofouling*, **22**, 339-360.

7. Falde, E. J., Yohe, S. T., Colson, Y. L., and Grinstaff, M. W. (2016) Superhydrophobic materials for biomedical applications. *Biomaterials*, **104**, 87-103.

8. Jin, M., Feng, X., Xi, J., Zhai, J., Cho, K., Feng, L., and Jiang, L. (2005) Super-hydrophobic PDMS surfaces with ultra-low adhesive force. *Macromolecular Rapid Communications*, **26**, 1805-1809.

9. Sun, M., Luo. C., Xu, L., Ji, H., Ouyang, Q., Yu, D. and Chen, Y. (2005) Artificial lotus leaf by nanocasting. *Langmuir*, **21**, 5549-5554.

10. Khorasani, M. T., Mirzadeh, H., and Kermani, Z. (2005) Wettability of porous polydimethylsiloxane surface: morphology study. *Applied Surface Science*, **242**, 339-345.

11. Bokobza, L. (2004) Elastomeric composites. I. Silicone composites. *Journal of Applied Polymer Science*, **93**, 2095-2104.

12. Chen, H., Zheng, M., Sun, H., and Jia, Q. (2007) Characterization and properties of sepiolite/polyurethane nanocomposites. *Materials Science and Engineering: A*, **445**, 725-730.

13. Ma, J., Bilotti, E., Peijs, T., and Darr J. A. (2007) Preparation of polypropylene/sepiolite nanocomposites using supercritical CO_2 assisted mixing. *European Polymer Journal*, **43**, 4931-4939.

14. Beigbeder, A., Degee, P., Conlan, S. L., Mutton, R. J., Clare, A. S., Pettitt, M. E., Callow, M. E., Callow, J. A., and Dubois, P. (2008) Preparation and characterization of silicone-based coatings filled with carbon nanotubes and natural sepiolite and their application as marine fouling release coatings. *Biofouling*, **24**, 291-302.

15. Wan, F., Pei, X., Yu, B., Ye, Q., Zhou, F., and Xue, Q. (2012) Grafting polymer brushes on biomimetic structural surfaces for anti-algae fouling and foul release. *ACS Applied Materials & Interfaces*, **4**, 4557-4565.

16. (a) Gittens, J. E., Smith, T. J., Suleiman, R., and Akid, R. (2013) Current and emerging environmentally-friendly systems for fouling control in the marine environment. *Biotechnology Advances*, **31**, 1738-1753; (b) Mohr, S., Berghahn, R., Mailahn, W., Schmiediche, R., Feibicke, M., and Schmidt, R. (2009) Toxic and accumulative potential of the antifouling biocide and TBT successor Irgarol on freshwater macrophytes - a pond mesocosm study. *Environmental Science & Technology*, **43**, 6838-6843.

17. Bodkhe, R. B., Stafslien, S. H., Cilz, N., Daniels, J., Thompson, S. E. M., Callow, M. E., Callow, J. A., and Webster, D. C. (2012) Polyurethanes with amphiphilic surfaces made using telechelic functional PDMS having orthogonal acid functional groups. *Progress in Organic Coatings*, **75**(1-2), 38-48.

18. Detty, M. R., Ciriminna, R., Bright, F. V., and Pagliaro, M. (2014) Environmentally benign sol-gel antifouling and foul-releasing coat-

ings. *Accounts of Chemical Research*, **47**, 678-687.

19. Ciriminna, R., Bright, F. V., and Pagliaro, M. (2015) Ecofriendly anti-fouling marine coatings. *ACS Sustainable Chemistry & Engineering*, **3**, 559-565.

20. (a) Howell, C., Vu, T. L., Lin, J. J., Kolle, S., Juthani, N., Watson, E., Weaver, J. C., Alvarenga, J., and Aizenberg, J. (2014) Self-replenishing vascularized fouling-release surfaces. *ACS Applied Materials & Interfaces*, **6**, 13299-13307; (b) Thomas, K. V., and Brooks, S. (2010) The environmental fate and effects of antifouling paint biocides. *Biofouling*, **26**, 73-88.

21. Brady, R. F., and Singer, I. S. (2000) Mechanical factors favoring release from fouling release coatings. *Biofouling*, **15**, 73-81.

22. (a) Rahaman, M. S., Therien-Aubin, H., Ben-Sasson, M., Ober, C. K., Nielsen, M., and Elimelech, M. (2014) Control of biofouling on reverse osmosis polyamide membranes modified with biocidal nanoparticles and antifouling polymer brushes. *Journal of Material Chemistry B*, **2**, 1724-1732; (b) Lejars, M., Margaillan, A., and Bressy, C. (2013) Well-defined graft copolymers of tert-butyldimethylsilyl methacrylate and poly(dimethylsiloxane) macromonomers synthesized by RAFT polymerization. *Polymer Chemistry*, **4**, 3282-3292.

23. Sankar G. G., Sathya S., Murthy P. S., Das A., Pandiyan R., Venugopalan V. P., and Doble, M. (2015) Polydimethyl siloxane nanocomposites: Their antifouling efficacy in vitro and in marine conditions. *International Biodeterioration & Biodegradation*, **104**, 307-314.

24. Yilgor, E., and Yilgor, I. (2014) Silicone containing copolymers: synthesis, properties and applications. *Progress in Polymer Science*, **39**, 1165-1195.

25. Satheesh, S., Ba-akdah, M. A., and Al-Sofyani, A. A. (2016) Natural antifouling compound production by microbes associated with marine macroorganisms - A review. *Electronic Journal of Biotechnology*, **21**, 26-35.

26. (a) Kango, S., Kalia, S., Celli, A., Njuguna, J., Habibi, Y., and Kumar, R. (2013) Surface modification of inorganic nanoparticles for development of organic-inorganic nanocomposites - A review. *Progress in Polymer Science*, **38**, 1232-1261; (b) Kuchibhatla, S. V. N. T., Karakoti, A. S., Bera, D., and Seal, S. (2007) One dimensional nanostructured materials. *Progress in Material Science*, **52**, 699-913.

27. (a) Jeon, I. Y., and Baek, J. B. (2010) Nanocomposites derived from polymers and inorganic nanoparticles. *Materials*, **3**, 3654-3674; (b) Taguet, A., Cassagnau, P., and Lopez-Cuesta, J. M. (2014) Structuration, selective dispersion and compatibilizing effect of (nano) fillers in polymer blends. *Progress in Polymer Science*, **39**, 1526-1563; (c) Karger-Kocsis, J., Mahmood, H., and Pegoretti, A. (2015) Recent advances in fibre/matrix interphase engineering for polymer composites. *Progress in Polymer Science*, **73**, 1-43; (d) Dang, Z.-M., Yuan,

J.-K., Zha, J.-W., Zhou, T., Li, S.-T., and Hu, G.-H. (2012) Fundamentals, processes and applications of high-permittivity polymer-matrix composites. *Progress in Material Science*, **57**, 660-723.

28. Nurioglu, A. G., Esteves, A. C. C., and de With, G. (2015) Non-toxic, non-biocide-release antifouling coatings based on molecular structure design for marine applications. *Journal of Material Chemistry B*, **3**, 6547-6570.

29. Cao, S., Wang, J. D., Chen, H. S., and Chen, D. R. (2011) Progress of marine biofouling and antifouling technologies. *Chinese Science Bulletin*, **56**, 598-612.

30. (a) Quigg, A., Chin, W. C., Chen, C. S., Zhang, S., Jiang, Y., and Miao, A. J., Kathleen, A., Schwehr, Xu, C., and Santschi, P. H. (2013) Direct and indirect toxic effects of engineered nanoparticles on algae: role of natural organic matter. *ACS Sustainable Chemistry & Engineering*, **1**, 686-702; (b) Baker, T. J., Tyler, C. R., and Galloway, T. S. (2014) Impacts of metal and metal oxide nanoparticles on marine organisms. *Environmental Pollution*, **186**, 257-271.

31. (a) Hussain, F., Hojjati, M., Okamoto, M., and Gorga R. E. (2006) Review article: polymer-matrix nanocomposites, processing, manufacturing, and application: an overview. *Journal of Composite Materials*, **40**, 1511-1575.

32. Owens, G. J., Singh, R. K., Foroutan, F., Alqaysi, M., Han, C. M., Mahapatra, C., Kim, H,-W., and Knowles, J. C. (2016) Sol-gel based materials for biomedical applications. *Progress in Material Science*, **77**, 1-79.

33. Lejars, M., Margaillan, A., and Bressy, C. (2012) Fouling release coatings: a non toxic alternative to biocidal antifouling coatings. *Chemical Reviews*, **112**, 4347-4390.

34. Yebra, D. M., Kiil, S., and Dam-Johansen, K. (2004) Antifouling technology past, present and future steps towards efficient and environmentally friendly antifouling coatings. *Progress in Organic Coatings*, **50**, 75-104.

35. Callow, J. A., and Callow, M. E. (2011) Trends in the development of environ-mentally friendly fouling-resistant marine coatings. *Nature Communications*, **2**, 244.

36. Chambers, L. D., Stokes, K. R., Walsh, F. C., and Wood, R. J. K (2006) Modern approaches to marine antifouling coatings. *Surface Coating Technology*, **201**, 3642-3652.

37. Almeida, E., Diamantino, T. C., and de Sousa, O. (2007) Marine paints: the particularcase of antifouling paints. *Progress in Organic Coatings*, **59**, 2-20.

38. Magin, C. M., Cooper, S. P., and Brennan, A. B. (2010) Non-toxic antifouling strategies. *Materials Today*, **13**, 36-44.

39. Rittschof, D. (2010) Research on practical environmentally benign antifouling coatings. In: *Biofouling*, Durr, S., and Thomason, J. C.

(eds.), Wiley-Blackwell, USA, pp. 396-409.

40. Clare, A. S., Rittschof, D., Gerhart, D. J., and Maki, J. S. (1992) Molecular approaches to non-toxic antifouling. *Invertebrate Reproduction & Development*, **22**, 67-76.

41. Baier, R. E., Shafrin, E. G., and Zisman, W. A. (1968) Adhesion - mechanisms that assist or impede it. *Science*, **162**, 1360-1368.

42. Roberts, D., Rittschof, D., Holm, E., and Schmidt, A. R. (1991) Factors influencing initial larval settlement - temporal, spatial and surface molecularcomponents. *Journal of Experimental Marine Biology and Ecology*, **150**, 203-221.

43. Baier, R. E., and Meyer, A. E. (1992) Surface analysis of fouling-resistant marine coatings. *Biofouling*. **6**,165-180.

44. Gerhart, D. J., Rittschof, D., Hooper, I. R., Eisenman, K., Mayer, A. E., Baier, R. E., and Young, C. (1992) Rapid and inexpensive quantification of the combinedpolar components of surface wettability: application to biofouling. *Biofouling*, **5**, 251-259.

45. Ista, L. K., Callow, M. E., Finlay, J. A., Coleman, S. E., Nolasco, A. C., Simons, R. H., Callow, J. A., and Lopez, G. P. (2004) Effect of substratum surface chemistry and surface energy on attachment of marine bacteria and algal spores. *Applied and Environmental Microbiology*, **70**, 4151-4157.

46. Bennett, S. M., Finlay, J. A., Gunari, N., Wells, D. D., Meyer, A. E., Walker, G. C., Callow, M. E., Callow, J. A., Bright, F. V., and Detty, M. R. (2010) The role of surface energy and water wettability in aminoalkyl/fluorocarbon/hydrocarbon-modified xerogel surfaces in the control of marine biofouling. *Biofouling*, **26**(2), 235-246.

47. Wu, L., Jasinski, J., and Krishnan, S. (2012) Carboxybetaine, sulfobetaine, andcationic block copolymer coatings: a comparison of the surface properties and antibiofouling behavior. *Journal of Applied Polymer Science*, **124**, 2154-2170.

48. Dahlstrom, M., Jonsson, H., Jonsson, P. R., and Elwing, H. (2004) Surface wettabilityas a determinant in the settlement of the barnacle Balanus impro-visus (DARWIN). *Journal of Experimental Marine Biology and Ecology*, **305**, 223-232.

49. Petrone, L., Di Fino, A., Aldred, N., Sukkaew, P., Ederth, T., Clare, A. S., and Lied-berg, B. (2011) Effects of surface charge and Gibbs surface energy on thesettlement behaviour of barnacle cyprids (Balanus amphitrite). *Biofouling*, **27**, 1043-1055.

50. Ma, J., Ma, C., Yang, Y., Xu, W., and Zhang, G. (2014) Biodegradable polyurethane carrying antifoulants for inhibition of marine biofouling. *Industrial & Engineering Chemistry Research*, **53**, 12753-12759.

51. Rasmussen, K., and Ostgaard, K. (2001) Adhesion of the marine fouling diatom Amphora coffeaeformis to non-solid gel surfaces. *Biofouling*, **17**, 103-115.

52. Rasmussen, K., Willemsen, P. R., and Ostgaard, K. (2012) Barnacle

settlement on hydrogels. *Biofouling*, **18**, 177-191.

53. Rasmussen, K., and Ostgaard, K. (2003) Adhesion of the marine bacterium Pseu-domonas sp NCIMB2021 to different hydrogel surfaces. *Water Research*, **37**, 519-524.

54. Evariste, E., Gatley, C. M., Detty, M. R., Callow, M. E., and Callow, J. A. (2013) The performance of aminoalkyl/fluorocarbon/hydrocarbon-modified xero gel coatings against the marine alga Ectocarpus crouaniorum: relativeroles of surface energy and charge. *Biofouling*, **29**, 171-184.

55. Ueshima, M., Tanaka, S., Nakamura, S., and Yamashita, K. (2002) Manipulation of bacterial adhesion and proliferation by surface charges of electrically polarized hydroxyapatite. *Journal of Biomedical Materials Research Part A*, **60**, 578-584.

56. Zhao, Y. H., Zhu, X. Y., Wee, K. H., and Bai, R. B. (2010) Achieving highly effective non-biofouling performance for polypropylene membranes modified by UV-induced surface graft polymerization of two oppositely charged monomers. *The Journal of Physical Chemistry B*, **114**, 2422-2429.

57. Kirschner, C. M., and Brennan, A. B. (2012) Bio-inspired antifouling strategies. *Annual Review of Materials Research*, **42**, 211-229.

58. Scardino, A. J., and de Nys, R. (2011) Mini review: biomimetic models and bioin-spired surfaces for fouling control. *Biofouling*, **27**, 73-86.

59. Bhushan, B., and Jung, Y. C. (2011) Natural and biomimetic artificial surfaces for super hydrophobicity, self-cleaning, low adhesion, and drag reduction. *Progress in Material Science*, **56**, 1-108.

60. Marmur, A. (2006) Super-hydrophobicity fundamentals: Implications to biofouling prevention. *Biofouling*, **22**, 107-115.

61. Salta, M., Wharton, J. A., Stoodley, P., Dennington, S. P., Goodes, L. R., Werwinski, S., Mart, U., Wood, R. J., and Stokes, K. R. (2010) Designing biomimeticantifouling surfaces. *Philosophical Transactions of the Royal Society A*, **368**, 4729-4754.

62. Efimenko, K., Finlay, J., Callow, M. E., Callow, J. A., and Genzer, J. (2009) Developmentand testing of hierarchically wrinkled coatings for marine antifouling. *ACS Applied Materials & Interfaces*, **1**, 1031-1040.

63. Aldred, N., and Clare, A. S. (2008) The adhesive strategies of cyprids and development of barnacle-resistant marine coatings. *Biofouling*, **24**, 351-363.

64. Zhao, Q., Su, X. J., Wang, S., Zhang, X. L., Navabpour, P., and Teer, D. (2009) Bacterial attachment and removal properties of silicon- and nitrogen-doped diamond-like carbon coatings. *Biofouling*, **25**, 377-385.

65. Baier, R. E. (2006) Surface behavior of biomaterials: the theta surface for biocompatibility. *Journal of Materials Science: Materials in*

Medicine, **17**, 1057-1062.

66. Yang, W. J., Neoh, K. G.,Kang, E. T., Teo, S. L. M., and Rittscho, D. (2014) Polymer brush coatings for combating marine biofouling. *Progress in Polymer Science*, **39**, 1017-1042.

67. Guo, Y. C., Yang, K., Zuo, X. H., Xue, Y., Marmorat C., Liu, Y., Chang, C. C., and Rafailovich, M. H. (2016) Effects of clay platelets and natural nanotubes on mechanical properties and gas permeability of Poly (lactic acid) nanocomposites. *Polymer*, **83**, 246-259.

68. Fischer, A. M., Schüll, C., and Frey, H. (2015) Hyperbranched poly(glycolide) copolymers with glycerol branching points via ring-opening copolymerization. *Polymer*, **72**, 436-446.

69. Ajiro, H., Takahashi, Y., Akashi, M., and Fujiwara, T. (2014) Surface control of hydrophilicity and degradability with block copolymers composed of lactide and cyclic carbonate bearing methoxyethoxyl groups. *Polymer*, **55**, 3591-3598.

70. Patil, N., Kelsey, J., Fischer, J., Grady, B., and White, J. L. (2014) Creating polymer templates and their use in the in-situ synthesis of biodegradable composite networks. *Polymer*, **55**, 2332-2339.

71. Fay, F., Linossier, I., Langlois, V., Renard, E., and Vallee-Rehel, K. (2006) Degradation and Controlled Release Behavior of ε-Caprolactone Copolymers in Biodegradable Antifouling Coatings. *Biomacromolecules*, **7**, 851-857.

72. Fay, F., Renard, E., Langlois, V., Linossier, I., and Vallee-Rehel, K. (2007) Development of poly(ε-caprolactone-co-l-lactide) and poly(ε-caprolactone-co-δ-valerolactone) as new degradable binder used for antifouling paint. *European Polymer Journal*, **43**, 4800-4813.

73. Fay, F., Linossier, I., Peron, J. J., Langlois, V., and Vallee-Rehel, K. (2007) Antifouling activity of marine paints: Study of erosion. *Progress in Organic Coatings*, **60**, 194-206.

74. Ma, C. F., Xu, L. G., Xu, W. T., and Zhang, G. Z. (2013) Degradable polyurethane for marine anti-biofouling. *Journal of Materials Chemistry B*, **1**, 3099-3106.

75. Xie, Q., Xie, Q., Pan, J., Ma, C., and Zhang, G. (2018) Biodegradable polymer with hydrolysis-iInduced zwitterions for antibiofouling. *ACS Applied Materials & Interfaces*, **10**(13), 11213-11220.

76. Xu, W. T., Ma, C. F., Ma, J. L., Gan, T. S., and Zhang, G. Z. (2014) Marine biofouling resistance of polyurethane with biodegradation and hydrolyzation. *ACS Applied Materials & Interfaces*, **6**, 4017-4024.

77. Yao, J. H., Chen, S. S., Ma, C. F., and Zhang, G. Z. (2014) Marine anti-biofouling system with poly (ε-caprolactone)/clay composite as carrier of organic antifoulant. *Journal of Materials Chemistry B*, **2**, 5100-5106.

78. Tachibana, Y., Masuda, T., Funabashi, M., and Kunioka, M. (2010) Chemical synthesis of fully biomass-based poly(butylene succin-

ate) from inedible-biomass-based furfural and evaluation of its bio-mass carbon ratio. *Biomacromolecules*, **11**, 2760-2765.

79. Liu, G. C., Zeng, J. B., Huang, C. L., Jiao, L., Wang, X. L., and Wang, Y. Z. (2013) Crystallization Kinetics and Spherulitic Morphologies of Bi-odegradable Poly(butylene succinate-*co*-diethylene glycol succin-ate) Copolymers. *Industrial & Engineering Chemistry Research*, **52**, 1591-1599.

80. Wu, B. S., Xu, Y. T., Bu, Z. Y., Wu, L. B., Li, B. G., and Dubois, P. (2014) Biobased poly(butylene 2,5-furandicarboxylate) and poly(butylene adipate-co-butylene 2,5-furandicarboxylate)s: From synthesis us-ing highly purified 2,5-furandicarboxylic acid to thermo-mechanical properties. *Polymer*, **55**, 3648-3655.

81. Ma, C., Xu, W., Pan, J., Xie, Q., and Zhang, G. (2016) Degradable poly-mers for marine antibiofouling: optimizing structure to improve performance. *Industrial & Engineering Chemistry Research*, **55**, 11495-11501.

82. Chen, S., Ma, C., and Zhang, G., (2016) Biodegradable polymers for marine antibiofouling: Poly(ε-caprolactone)/poly(butylene succin-ate) blend as controlled release system of organic antifoulant. *Poly-mer*, **90**, 215-221.

83. Movahedi, A., Zhang, J., Kann, N., Poulsen, K. M., and Nydén, M. (2016) Copper-coordinating polymers for marine anti-fouling coat-ings: A physicochemical and electrochemical study of ternary sys-tem of copper, PMMA and poly(TBTA). *Progress in Organic Coatings*, **97**, 216-221.

84. (a) Acevedo, M. S., Puentes, C., Carreno, K., Leon, J. G., Stupak, M., Garcia, M., Perez, M., and Blustein, G. (2013) Antifouling paints based on marine natural products from Colombian Caribbean. *In-ternational Biodeterioration & Biodegradation*, **83**, 97-104; (b) Srinivasan, M., and Swain, G. W. (2007) Managing the use of copper-based antifouling paints. *Environmental Management*, **39**, 423-441.

85. Misdan, N., Ismail, A. F., and Hilal, N. (2016) Recent advances in the development of (bio)fouling resistant thin film composite mem-branes for desalination. *Desalination*, **380**, 105-111.

86. (a) Higuchi, A., Sugiyama, K., Yoon, B. O., Sakurai, M., Hara, M., Su-mita, M., Sugawara, S., and Shirai, T. (2003) Serum protein adsorp-tion and platelet adhesion on pluronic (TM)-adsorbed polysul-phone membranes. *Biomaterials*, **24**, 3235-3245; Murosaki, T., Ah-med, N., and Gong, J. P. (2011) Antifouling properties of hydrogels. *Science and Technology of Advanced Materials*, **12**, 064706; (c) Lundberg, P., Bruin, A., Klijnstra, J. W., Nystrom, A. M., Johansson, M., Malkoch, M., and Hult, A. (2012) Poly(ethylene glycol)-based thiol-ene hydrogel coatings-curing chemistry, aqueous stability, and po-tential marine antifouling applications. *ACS Applied Materials & In-terfaces*, **2**, 903-912.

87. Higaki, Y., Kobayash, M., Murakami, D., Takahara, A. (2016) Anti-fouling behavior of polymer brush immobilized surfaces. *Polymer Journal*, **48**, 325-331.

88. (a) Satarkar, N. S., Biswal, D., Hilt, J. Z. (2010) Hydrogel nanocomposites: a review of applications as remote controlled biomaterials. *Soft Matter*, **6**, 2364-2371; (b) Chang, C. W., van Spreeuwel, A., Zhang, C., Varghese, S. (2010) PEG/clay nanocomposite hydrogel: a mechanically robust tissue engineering scaffold. *Soft Matter*, **6**, 5157-5164; (c) Kostina, N. Y., Sharifi, S., de los, A., Pereira, S., Michá-lek, J., Grijpma, D. W., and Emmenegger, C. R. (2013) Novel antifouling self-healing poly(carboxybetaine methacrylamide-co- HEMA) nanocomposite hydrogels with superior mechanical propertie. *Journal of Materials Chemistry B*, **1**, 5644-5650; (d) Wang, J, Hu, H, Yang, Z, Wei, J, and Li, J. (2016) IPN hydrogel nanocomposites based on agarose and ZnO with antifouling and bactericidal properties. *Materials Science and Engineering: C*, **61**, 376-386.

89. Kingshott, P., Thissen, H., and Griesser, H. J. (2002) Effects of cloud-point grafting chain length, and density of PEG layers on competitive adsorption of ocular proteins. *Biomaterials*, **23**, 2043-2056.

90. (a) Schilp, S., Rosenhahn, A., Pettitt, M. E., Bowen, J., Callow, M. E., Callow, J. A., Grunze, M. (2009) Physicochemical properties of (ethylene glycol)-containing self-assembled monolayers relevant for protein and algae cell resistance. *Langmuir*, **25**, 10077-10082; (b) Lowe, S., O'Brien-Simpson, N. M., and Connal, L. A. (2015) Antibiofouling polymer interfaces: poly(ethylene glycol) and other promising candidates. *Polymer Chemistry*, **6**, 198-212; (c) Pelaz, B., Del Pino, P., Maffre, P., Hartmann, R., Gallego, M., Rivera-Fernandez, S., de la Fuente, J. M., Nienhaus, G. U., and Parak, W. J. (2015) Surface functionalization of nanoparticles with polyethylene glycol: effects on protein adsorption and cellular uptake. *ACS Nano*, **7**, 6996-7008; (d) Wang, W. Y., Shi, J. Y., Wang, J. L., Li, Y. L., Gao, N. N., Liu, Z. X., and Lian, W. T. (2015) Preparation and characterization of PEG-g-MWCNTs/PSf nano-hybrid membranes with hydrophilicity and antifouling properties. *RSC Advances*, **5**, 84746-84753.

91. Huang, G. (2014) *Laser-Assisted Surface Modification of Hybrid Hydrogels to Prevent Bacterial Contamination and Protein Fouling*, Maters Thesis, University of Western Ontario, Canada.

92. Barua, S., Chattopadhyay, P., Phukan, M. M., Konwar, B. K., Islam, J., and Karak, N. (2014) Biocompatible hyperbranched epoxy/silver-reduced graphene oxide-curcumin nanocomposite as an advanced antimicrobial material. *RSC Advances*, **4**, 47797-47805.

93. (a) Barua, S., Dutta, G., and Karak, N. (2013) Glycerol based tough hyperbranched epoxy: synthesis, statistical optimization and property evaluation. *Chemical Engineering Science*, **95**, 138-147; De, B., and Karak, N. (2013) Novel high performance tough hyperbranched

epoxy by an A2 + B3 polycondensation reaction. *Journal of Materials Chemistry A*, **1**, 348-353; (c) Suriyanarayanan, S., Lee, H. H., Liedberg, B., Aastrup, T., and Nicholls, I. A. (2013) Protein- resistant hyperbranched polyethyleneimine brush surfaces. *Journal of Colloid and Interface Science*, **396**, 307-315.

94. (a) Zhou, C., Shi, Y., Sun, C., Yu, S., Liu, M., Gao, C. (2014) Thin-film composite membranes formed by interfacial polymerization with natural material sericin and trimesoyl chloride for nanofiltration. *Journal of Membrane Science*, **471**, 381-391; (b) Deka, H., Karak, N., Kalita, R. D., Buragohain, A. K. (2010) Bio-based thermostable, biodegradable and biocompatible hyperbranched polyurethane/Ag nanocomposites with antimicrobial activity. *Polymer Degradation and Stability*, **95**, 1509-1517; (c) Wei, Q., Krysiak, S., Achazi, K., Becherer, T., Noeskec, P. L. M., Paulus, F., Liebe, H., Grunwald, I., Dernedde, J., Hartwig, A., Hugel, T., and Haag, R. (2014) Multivalent anchored and cross linked hyper branched polyglycerol monolayers as antifouling coating for titanium oxide surfaces. *Colloids and Surfaces B: Biointerfaces*, **122**, 684-692.

95. Selim, M. S., Shenashen, M. A., El-Safty, S. A., Higazy, S.A., Selim, M. M., Isago, H., and Elmarakbi, A. (2017) Recent progress in marine foul-release polymeric nanocomposite coatings. *Progress in Materials Science*, **87**, 1-32.

96. (a) Meng, J., Cao, Z., Ni, L., Zhang, Y., Wang, X., Zhang, X. and Liu, E. (2014) A novel salt-responsive TFC RO membrane having superior antifouling and easy-cleaning properties. *Journal of Membrane Science*, **461**, 123-129; (b) Jiang, S. Y., and Cao, Z. Q. (2010) Ultralow-fouling functionalizable and hydrolysable zwitterionic materials and their derivatives for biological applications. *Advanced Materials*, **22**, 920-932.

97. (a) Zhang, Z., Finlay, J. A., Wang, L., Gao, Y., Callow, J. A., Callow, M. E., and Jiang, S. (2009) Polysulfobetaine-grafted surfaces as environmentally benign ultra low fouling marine coatings. *Langmuir*, **25**, 13516-13521; (b) Zhu, J., Zhao, X., and He, C. (2015) Zwitterionic SiO2 nanoparticles as novel additives to improve the antifouling properties of PVDF membranes. *RSC Advances*, **5**, 53653-53659; (c) Li, Y., Giesbers, M., Gerth, M., and Zuilhof, H. (2012) Generic top-functionalization of patterned antifouling zwitterionic polymers on indium tin oxide. *Langmuir*, **28**, 12509-12517; (d) Jeong, C. J., Kang, E. B., Park, S. J., Choi, K. H., Shin, G., In, I., and Park, S. Y. (2015) Preparation of exfoliated montmorillonite nanocomposites with catechol/zwitterionic quaternized polymer for an antifouling coating. *Polymer Engineering & Science*, **55**, 2111-2117.

98. (a) Lung, K. Y., and Dario, N. D. (2015) Surface modification strategies on mesoporous silica nanoparticles for anti-biofouling zwitterionic film grafting. *Advances in Colloid and Interface Science*, **226**,

166-186; (b) Hu, R., Li, G., Jiang, Y., Zhang, Y., Zou, J.-J., Wang, L., and Zhang, X. (2013) Silver-zwitterion organic-inorganic nanocomposite with antimicrobial and antiadhesive capabilities. *Langmuir*, **29**, 3773-3779.

99. Wang, J., and Wei, J. (2016) Hydrogel brushes grafted from stainless steel via surface-initiated atom transfer radical polymerization for marine antifouling. *Applied Surface Science*, **382**, 202-216.

100. Chen, S., Ma, C., and Zhang, G. (2017) Biodegradable polymer as controlled release system of organic antifoulant to prevent marine biofouling. *Progress in Organic Coatings*, **104**, 58-63.

101. Shi, J., Wanga, X., Wu, Z., and Zhu, Z. (2017) Fatigue behavior of basalt fiber-reinforced polymer tendons under a marine environment. *Construction and Building Materials*, **137**, 46-54.

102. Yesudass, S. A., Mohanty, S., Nayak, S. K., and Rath, C. C. (2017) Zwitterionic-polyurethane coatings for non-specific marine bacterial inhibition: A nontoxic approach for marine application. *European Polymer Journal*, **96**, 304-315.

103. Burgess, J. G., Boyd, K. G., Armstrong, E., Jiang, Z., Yan, L., Berggren, M., May, U., Pisacane, T., Granmo, A., and Adams, D. R. (2003) The development of a marine natural product-based antifouling paint. *Biofouling*, **19**, 197-205.

104. Qian, P. Y., Lau, S. C. K., Dahms, H. -U., Dobretsov, S., and Harder, T. (2007) Marine biofilms as mediators of colonization by marine macro organisms: implications for antifouling and aquaculture. *Marine Biotechnology*, **9**, 399-410.

105. Salta, M., Wharton, J. A., Blache, Y., Stokes, K. R., Briand, J. F. (2013) Marine biofilms on artificial surfaces: structure and dynamics. *Environmental Microbiology*, **15**, 1879-1893.

106. Naamani, L. A., Dobretsov, S., Dutta, J., and Burgess, J. G., (2017) Chitosan- zinc oxide nanocomposite coatings for the prevention of marine biofouling. *Chemosphere*, **168**, 408-417.

107. Yee, M. S.-L., Khiew, P.-S., Chiu, W. S., Tan, Y. F., Kok, Y.Y., and Leong, C.-O. (2016) Green synthesis of graphene-silver nanocomposites and its application as a potent marine antifouling agent. *Colloids and Surfaces B: Bio nterfaces*, **148**, 392-401.

108. Sathya, S., Murthy, P. S., Das, A., Sankar, G. G., Venkatnarayanan, S., Pandian, R., Sathyaseelan, V. S., Pandiyan, V., Doble, M., and Venugopalan, V. P. (2016) Marine antifouling property of PMMA nanocomposite films: Results of laboratory and field assessment. *International Biodeterioration & Biodegradation*, **114**, 57-66.

Recent Advances in Designed Non-toxic Polysiloxane Coatings to Combat Marine Biofouling

Elisa Martinelli,* Elisa Guazzelli and Giancarlo Galli

Dipartimento di Chimica e Chimica Industriale, Università di Pisa, Via G. Moruzzi 13, 56124 Pisa, Italy

Corresponding author: elisa.martinelli@unipi.it

6.1 Introduction

Marine biofouling is a prime example of a naturally-occurring nanoscale adhesion process, involving interfacial interactions between the submerged surfaces and the specific structural elements of the adhesives of the marine organisms, such as microorganisms (bacteria, diatoms) and macroorganisms, including soft algae, sponges and calcareous macrofoulers. It is a global practical and economic problem that has detrimental effects on all submerged surfaces, e.g., shipping and leisure crafts, aquaculture systems, oceanographic sensors and heat exchanger tubes. Heavily fouled ship hulls were reported to suffer from a reduction in cruising speed and an increase in fuel consumption with a powering penalty of up to 86% at cruising speed; slimes can also lead to a significant increase in powering penalty of 10-16% [1]. Beside the significant economic impact of biofouling on shipping [2], environmental issues associated with the increase in fuel consumption and greenhouse gas emissions, dry docking operations and transportation of invasive alien species in a non-native environment are also relevant issues [3-5].

Since the 1970s, marine biofouling was successfully controlled by the use of self- polishing copolymer (SPC) antifouling paints containing tributyltin (TBT), a broad spectrum biocide, offering ship operators cost-effective 60+ month foul-free hulls which resulted in significant reductions in fuel consumption [6]. However, the negative effects of TBT [7] forced many governments to restrict its use, and its ban came in force in 2008 [8]. The tin-based coatings were sub-

Marine Coatings and Membranes, edited by Vikas Mittal
© 2019 Central West Publishing, Australia

sequently replaced by more environmentally-friendly SPCs containing a mixture of cuprous oxide and organic booster biocides, e.g., zinc and copper pyrithione (Figure 6.1) [9]. Such TBT-free products are still used by the majority of the world-wide fleet. These coatings are becoming progressively more regulated because of the environmental concerns due to the high level of copper leaching, negatively affecting different marine organisms at various stages of their life [10]. Over the last years, non-toxic and biocide free [11-13] antifouling technologies have gained a great deal of interest. Generally, 'green' technologies rely on surface physico-chemical and bulk material properties either to prevent the initial attachment of micro- and macro-foulers according to an antifouling (AF) mechanism, and/or to promote the easy detachment of marine organisms, eventually attached by weak bonds, under low shear stresses (i.e. the hydrodynamic forces generated by the sailing of the vessel or the weight of the foulers themselves), thus, acting according to a fouling-release (FR) mechanism [14]. However, the mere distinction in AF and FR approaches is over-simplistic, as these two mechanisms are not mutually exclusive and often cooperate in the same experimental coating.

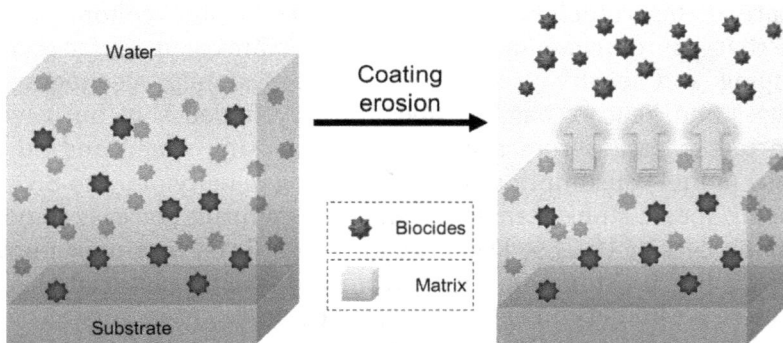

Figure 6.1 Schematic illustration of a biocide-based self-polishing copolymer (SPC) coating.

The most promising fouling-resistant coatings for application to large-scale structures, such as ship hulls, are based on polydimethylsiloxane (PDMS) binders that mainly act according to a FR strategy. PDMS-based FR coatings were recognized as a valid eco-friendly answer to fill the technological gap left by the legislation restrictions on the use of biocide-based AF coatings [6,15].

Starting from this rationale, a general description of the basic parameters of a FR coating will be provided in this chapter. The physico-chemical properties of PDMS will be revised and discussed in view of its application as a base for FR systems, by encompassing the most relevant strategies developed to improve FR as well as mechanical properties and durability in the marine environment of the coating. Particular attention will be devoted to the description of the latest approaches for the chemical modification of the hydrophobic PDMS to render the surface more hydrophilic and the entire coating amphiphilic, with an aim to enhance its AF activity as well as original FR capacity against microfoulers and to improve its performance when applied to vessels that cruise at relatively low speeds (<15 knots) or spend idle periods in port. Finally, a paragraph will be devoted to the description of silicone-based technologies currently commercialized by leading companies in the field of marine antifouling coatings.

6.2 Fouling-release Coatings (FRCs)

Several physico-chemical and mechanical parameters are known to affect the FR efficacy of a coating, including surface energy, elastic modulus and film thickness [16]. In particular, an empirical relationship, known as the Baier curve, between the relative adhesion of fouling organisms and the critical surface tension of the substrate shows that minimal fouling does not occur for the minimum value of surface energy, but for values in the range 22-24 mN/m, typical of poly(dimethylsiloxane) (PDMS) [17]. Indeed, the relative adhesion was demonstrated to linearly correlate with the square root of the product of the Young elastic modulus (E) and the critical surface tension (γ_c) with the smallest adhesion being achieved for soft and low elastic modulus materials, i.e., PDMS elastomers (Figure 6.2) [18]. The higher macromolecular mobility of low-modulus polymers allows the adhesive to slip during interfacial failure, reducing the energy input needed to achieve failure. The mechanism for the removal of biofoulers from soft films is understood [19-21], at least qualitatively, on the basis of Kendall's studies [22], showing that the pull-off force, P_c, required to remove a rigid cylindrical stud of diameter a from an elastomeric glue film of thickness t, when $a \gg t$, is proportional to $(WE/t)^{1/2}$, where W is the work of adhesion between the cylinder and the elastomer. Practically, it was shown that the releases of both *Ulva linza* sporelings and *Balanus amphitrite* (= *Am-*

phibalanus amphitrite) [23] adults were directly proportional to the thickness of the PDMS coating [24,25]. Therefore, PDMS-based elastomers, which combine together the low surface energy with the low elastic modulus, were considered as the best promise biocidefree alternatives.

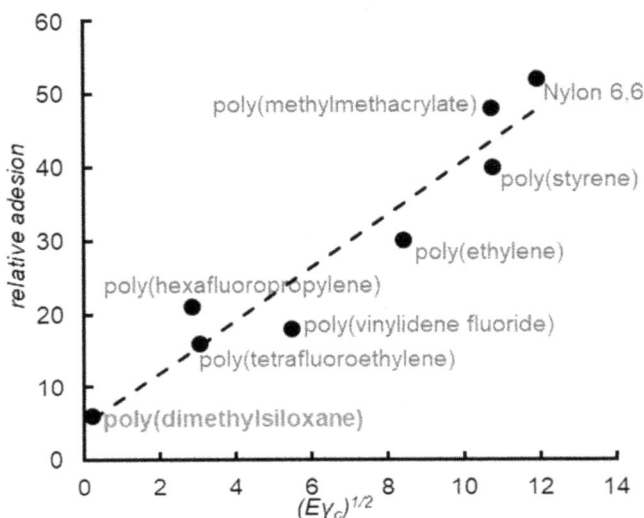

Figure 6.2 Relationship between the relative adhesion and the elastic modulus and the critical surface tension.

6.3 Polysiloxanes-based FRCs

Despite polysiloxane-based coatings started to be developed at the same time as SPCs, they were first applied on a small boat in 1987 [26].

PDMS is a hybrid organic-inorganic polymer composed of an inorganic backbone of the repeating Si–O unit and two organic CH_3 groups attached to the Si atom (Figure 6.3) [27]. Due to the electronegativity difference between Si and O, the Si–O bonds are strongly polarized, highly ionic and have a large bond energy (up to 460 kJ/mol). However, the side methyl groups shield the main chain, thus preventing strong intermolecular interactions. Moreover, the very weak interactions among the polysiloxane chains with the methyl group pointing to the outside are responsible for the high hydrophobicity and low surface tension of PDMS surfaces. However, a

decrease in the hydrophobic nature of PDMS upon relatively prolonged contact with water is also reported, due to surface rearrangement phenomena, which involve the migration to and away
from the polymer–water interface of the Si–O groups and the CH_3
groups, respectively [28].

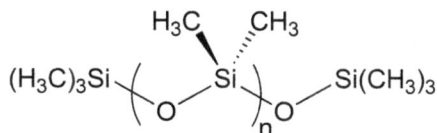

Figure 6.3 Chemical structure of PDMS.

PDMS is also characterized by low elastic modulus and low surface roughness, which are both known to drastically decrease the
adhesion strength of biofoulers. Rotation around Si–O bonds is vir-
tually free, the energy being almost zero, compared to 14 J/mol for
rotation about C–C bonds in polyethylene and 20 J/mol for
poly(tetrafluoroethylene). This confers to PDMS high flexibility and
mobility, which are reflected in the very low glass transition temperature (T_g = -127 °C). A low T_g was suggested to minimize mechanical interlocking between the biofoulers and the PDMS surface
[29,30].

The polysiloxanes used in the FR coatings are polymer elastomers normally obtained by one of two different reactions: hydrosilylation and polycondensation [16]. In any case, the coating film is
composed of a PDMS-based matrix, a crosslinker and a catalyst.

The hydrosilylation reaction occurs between two precursors of
polysiloxane type: one comprising vinyl end-groups and the other
bearing hydrosilane groups, which act as crosslinkers (Figure 6.4a).
Various types of platinum catalysts are typically used in hydrosilylation reactions [16]. One major drawback in this reaction system de-
rives from the potential toxicity of the platinum catalyst. In the condensation reaction, an α,ω-bissilanol-terminated PDMS and a poly-
acetoxy or polyalkoxy silane crosslinker react upon contact with
moisture of air, normally in the presence of a tin catalyst [16] (Figure 6.4b). Bismuth catalysts have also been used [31]. In all of these
cases, low molecular weight acids or alcohols are formed as byprod-
ucts.

The incorporation of relatively low molecular weight silicones into the siloxane network, as non-reactive surface-active additives of-

ten referred to as silicone oils, was widely reported to enhance the FR properties of the PDMS polymers [33-36]. By segregating to the polymer surface, the oil acts as wetting agent modifier, thus, affecting the hydrophobicity and surface energy of the matrix during curing and after water immersion, and as a lubricant, thus, reducing the coefficient of friction and allowing the detachment of the foulers by slippage.

(a) Hydrosilylation

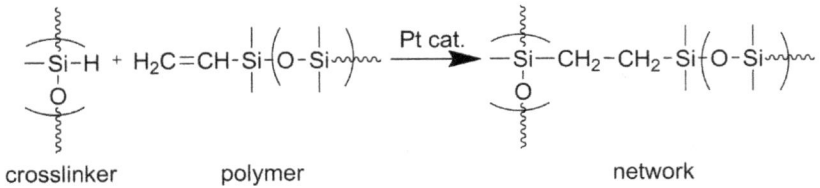

(b) Condensation

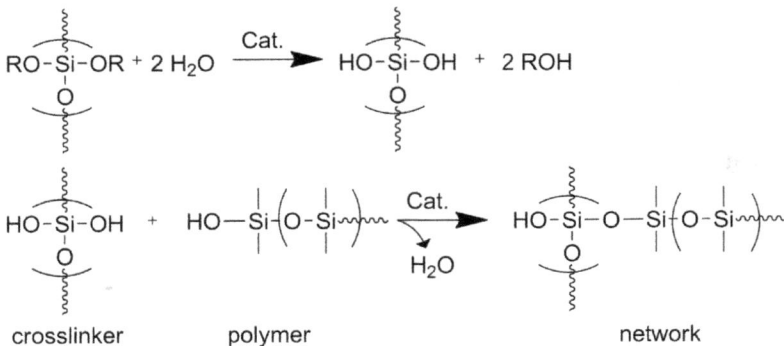

Figure 6.4 Hydrosilylation (a) and condensation (b) curing reactions of PDMS.

Environmental concerns associated with PDMS-derived FRCs seem to be minor with respect to those of copper-based SPCs and are related to possible leaching and persistence in seawater of non-reactive oils, un-crosslinked polysiloxane chains and curing catalysts [16].

Despite the desired combination of low surface energy and low elastic modulus as well as their efficacy in promoting the release of macrofoulers under hydrodynamic shear stresses, traditional PDMS-based coatings suffer from several drawbacks. They are difficult to adhere to a substrate without an appropriate tie coat and have poor mechanical properties that make them less durable and more easily damaged than other coatings [16]. In addition, they are poorly effective against slimes, dominated by diatoms, which are known to adhere more strongly to hydrophobic surfaces [37-39]. Finally, they are more effective at relatively high cruising speeds and less suitable in static conditions, i.e. for ships that spend long periods in port [14].

6.4 Chemically Modified PDMS-based FRCs

In order to improve the mechanical properties, adhesion to substrate, underwater durability and stability of PDMS-based FRCs, several strategies have been proposed, including the incorporation of inorganic (nano)fillers, modification of polysiloxane matrix with polyurethane or epoxy resins and addition of fluoropolymers.

Mechanical reinforcement of PDMS elastomers was achieved by incorporating inorganic fillers, such as calcium carbonate, silica and titania. However, the incorporation of these fillers may reduce the hydrophobicity of the PDMS matrix, thus, promoting the adhesion of the foulers proportionally to the amount of fillers [13,40,41]. Interesting insights were obtained by Beigbeder *et al.* [42] by dispersing natural sepiolite nanofibers in the PDMS matrix. A significant increase in the tensile modulus from 2.6 MPa for the unfilled formulation to 6 MPa for the 10 wt% sepiolite formulation was observed. Differently, the mechanical properties of PDMS were not improved by addition of a low amount of multiwall carbon nanotubes (MWCNs) ($\leq$ 0.2 wt%), while the increment in PDMS matrix wettability (i.e., loss of hydrophobicity) was comparatively low upon immersion in water for coatings loaded with 0.1 wt% MWCNs [43]. Interestingly, it was found that the formulation containing 0.05 wt% MWCNs improved the release of *U. linza* sporelings and zoospores as well as halved the critical removal stress of *B. amphitrite* adults [42].

Self-stratified polysiloxane-polyurethanes and polysiloxane-epoxy resins resulted in PDMS-based coatings with improved mechanical robustness, durability and adhesion to substrate [16]. Due to the chemical incompatibility between polysiloxane on one hand and the polyurethane (PU) or epoxy resin on the other hand, a phase separation of the two components in the formulation occurs that results in a self-stratification of the coating with the lowest surface energy polysiloxane component being preferentially segregated at the topmost surface. The self-stratified layer of PDMS imparts FR properties to the films comparable to those of commercial FR coatings, while the polyurethane bulk provides superior mechanical properties and adhesion to primers, thus, avoiding the use of a tie coat [44-46].

FR polysiloxane coatings incorporating hydrophobic reactive surface-active additives (see 6.5.2.), such as functionalized fluorinated polymers, were developed in an attempt to combine in one single structure the key features of both polysiloxanes and fluoropolymers, namely their low surface energy, surface roughness and elastic modulus. Grunlan and coworkers [47] cured a diaminoterminated PDMS matrix with star oligofluorosiloxanes terminated with epoxy groups in order to reduce the settlement and promote the release of algae and barnacles (Figure 6.5). In another example, a fluorinated trialkoxysilane was used as crosslinker for the polycondensation reaction of an α,ω-dihydroxysiloxane (Figure 6.5) [48]. The surface enrichment in the lowest surface energy fluorinated component ensured a longer stability of the coatings when exposed to water with respect to the non-fluorinated PDMS network. Incorporation of hydrophobic non-reactive surface-active additives (see 6.5.1.) is also reported as a strategy to improve the FR properties and stability of PDMS in water [49]. As an example,

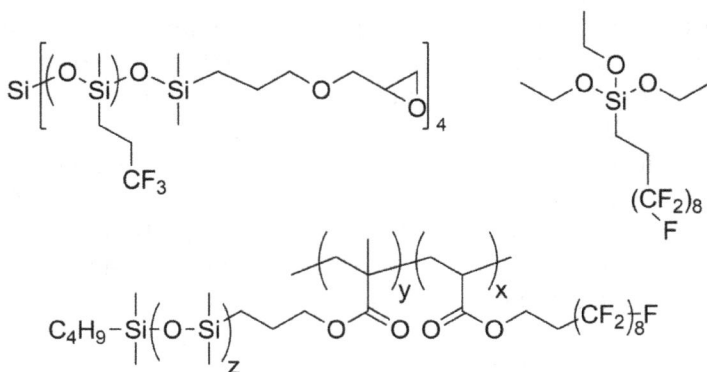

Figure 6.5 Illustrations of reactive (top) [47,48] and non-reactive (bottom) [50] surface-active fluorinated additives for incorporation into a PDMS matrix network.

Marabotti *et al.* [50] prepared blends of a fluorinated/siloxane copolymer and a condensation curing PDMS matrix (Figure 6.5). Film surfaces were highly enriched in the fluorinated component and were able to reduce the settlement of *B. amphitrite* cyprids in a co-polymer concentration-dependent manner. The coatings also exhibited higher release performance of *U. linza* sporelings and adult barnacles than pure PDMS control. Moreover, the non-reactive fluorinated/siloxane surface-active copolymer was modified into a reactive additive by incorporation of 3-(trimethoxysilyl)propyl methacrylate as a condensation-curable comonomer, able to act as a cross-linker in the curing reaction of dihydroxy-terminated PDMS matrix. The chemical anchoring of the polymer allowed for a better stabilization of the fluorine-enriched surface layer and prevention of possible leaching out of the copolymer [51]. Coating formulations containing such a copolymer displayed lower adhesion strength of pseudobarnacles than fluorine-free PDMS FRCs [51].

6.5 Hydrophilization of PDMS-based Coatings with Improved AF/FR Performance

One of the main pursued strategies to enhance the fouling resistance of polysiloxane coatings is to render their surface amphiphilic [52]. Amphiphilic polymers can create a heterogeneous nanoscale mosaic chemical surface, where the simultaneous presence of hydrophobic and hydrophilic domains is expected to make the surface "ambiguous" and unsuitable for the attachment of proteins, cells and marine micro- and macro-organisms [53,54] (Figure 6.6). Different organisms are known to interact with the surface according to specific adhesion profiles and contrasting preferences. As an example, the diatom *Navicula perminuta* and the macroalga *U. linza* adhere preferentially on hydrophobic and hydrophilic substrates, respectively

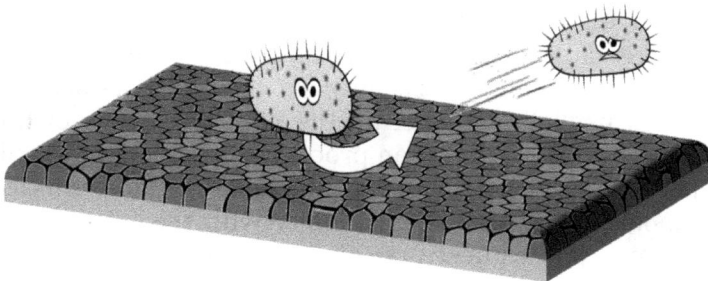

Figure 6.6 Illustration of the heterogeneous surface structure of an amphiphilic polymer with antifouling properties.

[14]. Moreover, for chemically heterogeneous coatings in general and amphiphilic surfaces in particular, antifouling properties are strongly affected by the length scale of different chemical functionalities at the surface; for example, the critical length scale for the settlement of *U. linza* zoospore is on the micrometer scale [55]. However, achieving the optimum hydrophilic/hydrophobic balance is still a demanding challenge. Among the diverse amphiphilic polymer platforms [56], PDMS-based amphiphilic coatings are one of the most promising strategies to be further investigated on a laboratory-scale and implemented for a large-scale industrial production.

Fouling resistant amphiphilic siloxane coatings were developed by different physical or chemical surface functionalization approaches, including:

- incorporation of non-reactive, surface-active amphiphilic polymer additives into a polysiloxane matrix;
- incorporation of reactive, surface-active amphiphilic additives into a polysiloxane matrix;
- hydrophilization of self-stratified siloxane-polyurethane coatings;
- zwitterionization of polysiloxane surfaces;
- modification of polysiloxanes with biomolecules and biomimetic molecules.

Selected examples of each category will be discussed in the following sections, especially in view of their potential application as marine AF/FR coatings.

6.5.1 Incorporation of Amphiphilic Non-reactive, Surface-active Polymer Additives

A non-reactive surface-active polymer is a physically dispersed additive capable of segregating and accumulating at the surface of the matrix, thereby modifying the physico-chemical character of the surface with negligible impact on the bulk properties of the film [57,58]. The surface segregation process of surface-active additives was reported to depend on several factors, including surface free energy, molecular weight, compatibility with the host matrix and enthalpy gain upon contact with the external environment [50,59-61]. Typical non-reactive additives consist of different components that provide complementary properties. Generally, a constituent structurally similar to the matrix is introduced in order to impart compatibility with the binder to prevent macrophase separation and delamination of the additive. The other components are chosen to provide the desired antifouling properties strictly dependent on the surface properties of the additives, such as surface wettability, composition, morphology, topography, reconstruction upon immersion in water and lubricity.

PDMS is one of the most common binders for the dispersion of the surface-active additives, in order to combine the bulk (elastomeric) properties of the matrix with the surface properties of the additives. The preparation of the films can be performed according to one-layer or two-layer geometries (Figure 6.7). In any case, maximal amounts of the additive do not exceed 10 wt%. However, the latter geometry was reported to allow for a better control of the bulk thickness and elastic modulus on one hand, and the surface and interface properties on the other [62].

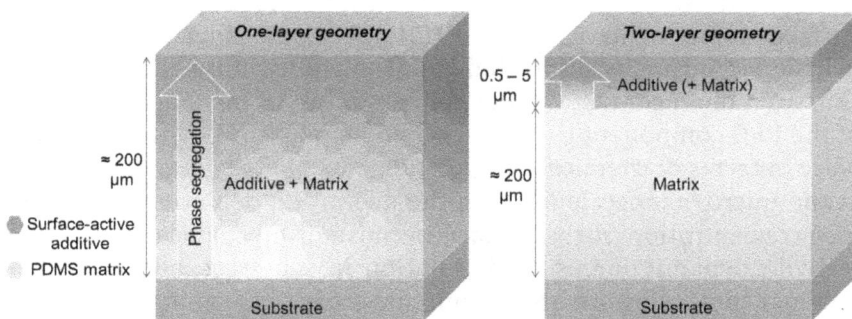

Figure 6.7 Schematic illustration of one-layer and two-layer geometries of PDMS-based coatings.

In addition to PDMS, poly(styrene-*b*-(ethylene-*co*-butene)-*b*-styrene) (SEBS) was also used as an elastomeric matrix for the preparation of one-layer or two-layer films containing hydrophobic or amphiphilic surface-active copolymers for AF/FR applications in the marine environment [60,62,63-65].

Amphiphilic additives are a new generation of surface-active modifiers which generally contain poly(ethylene glycol) (PEG) as the hydrophilic component covalently linked to the hydrophobic polysiloxane and/or fluorinated components [66]. PEG is a linear, non-polar and hydrophilic polymer with a remarkable resistance to the attachment of proteins, bacteria, cells and marine organisms. Although PEG-coated surfaces have a relatively high surface energy

(>43 mN/m), they possess a very low interfacial energy (~5 mN/m) [12] with water, and the resistance to protein adsorption is reported to rely on the steric repulsion between proteins and hydrated neutral PEG chains, which form a strong hydration layer with water through hydrogen bonds [67,68].

Several amphiphilic copolymers with different macromolecular architectures were developed in the last 10 years as additives for PDMS matrix for AF/FR applications. Some examples are provided in Table 6.1. In any case, the copolymers consist of a siloxane segment to improve the chemical compatibility with the PDMS matrix and a hydrophilic PEG segment to impart amphiphilic properties to the system and responsiveness to the external environment. A fluorinated component directly linked to or separated from the PEG segment can also be attached to enhance the low surface energy properties of the copolymer and its ability to migrate to the polymer–air interface, thus, concentrating the PEG chains close to the outermost surface layers. Generally, the AF/FR effectiveness against different marine organisms is associated with the ability of the surface to reconstruct upon contact with water as a result of the major exposure of the PEG component to water and an increase in surface heterogeneity in terms of chemical composition, morphology and topography at the micro/nanoscale level. In the event of a depletion in copolymer concentration at the surface because of mechanical stress due to hydrodynamic forces or dissolution in seawater, additional copolymer molecules are able to migrate from the film bulk to the interface to replace those washed away, in a sort of self-healing/self-repairing process. This process is effective until fresh copolymer is available in the bulk of the film. However, the release of the copolymer should proceed at low speed to guarantee a long-term application.

6.5.2 Incorporation of Amphiphilic Reactive Surface-active Additives

In order to overcome possible drawbacks associated with the leaching out of non-reactive additives from the polymer matrix after prolonged contact with water, efforts have been devoted to prepare PDMS-based amphiphilic coatings with covalently attached surface-modifying agents. Different hydrophilic and hydrophobic surface-active (meth)acrylic macromonomers having PEG and perfluoroalkyl grafts were employed either individually or in combination and

Table 6.1 Examples of amphiphilic copolymer structures used as non-reactive surface-active additives and their AF/FR performance

Amphiphilic copolymer structure	AF/FR performance
C$_4$H$_9$–Si(–O–Si)$_z$ ... O ... block copolymer with $(CF_2)_6$F and PEGylated side groups	Films with 1–4 wt% low molecular weight (19000 g/mol $\leq M_n \leq$ 31000 g/mol) block copolymer perform much better than PDMS control and exhibit high removal (up to 100% at 8 Pa) of *U. linza* sporelings. The composition ratios from 50:50 to 60:40 of the fluorinate/PEGylated side groups are shown to be more effective, with several films exhibiting spontaneous detachment of the sporelings. Performance against *Navicula incerta* is inferior, with some of the coatings having a percentage of removal as good as PDMS control [69].
H–Si(–O–Si)$_z$ ... O(–O–)$_x$	Adsorption of fibrinogen before and after 2 weeks of water conditioning is very low (< 15 ng/cm^2) with respect to the unmodified silicone (> 125 ng/cm^2). [70].
PDMS-PEG copolymers with the following structures: Copolymer 1: AB block, 8 PEG units in A, M_w = 660; Copolymer 2: AB graft, 10 PEG units in A, M_w = 3940; Copolymer 3: ABA block, 10 PEG units in A, M_w = 1960.	Coatings with 4 wt% copolymer show very good biofouling-resistance upon immersion in static conditions in seawater for 2 months. The biofouling community consists mainly of slime, with some tubeworms and bryozoans [71].
C$_4$H$_9$–Si(–O–Si)$_z$ ... copolymer with pentafluorostyrene and PEG-modified groups	Films containing 4 wt% copolymer are more effective in inhibiting the settlement of *B. amphitrite* than *B. improvisus* cyprids. Moreover, *B. improvisus* juveniles are more easily detached from films containing the copolymer with a lower amount of PEG-modified pentafluorostyrene [72,73].
C$_4$H$_9$–Si(–O–Si)$_z$ ... copolymer with pentafluorostyrene and PEG-modified groups	Films containing 8 wt% copolymer with the higher content of PEG-modified pentafluorostyrene show greater resistance to settlement of zoospores of *U. linza*, whereas all the films promote the release of *U. linza* sporelings better than the PDMS control, being the percentage removal up to ~ 60% at 13 Pa shear stress [74].
C$_4$H$_9$–Si(–O–Si)$_z$ ... copolymer with $(CF_2)_6$F and PEGylated counits	Films containing 1–4 wt% copolymer with lower amount of perfluorohexyl-PEGylated counits exhibit good FR properties against *U. linza* sporelings (up to ~ 60% removal at 8 Pa shear stress) [75]. However, shortening the perfuoroalkyl side chain results in a reduction of the FR performances compared to the structurally similar perfluorooctyl-PEGylated copolymers [76].

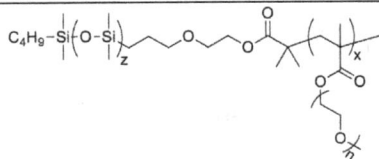

Structure	Description
$C_4H_9-Si(O-Si)_z$ ...	Films containing 4–8 wt% block copolymer in the top layer show a good removal of *B. amphitrite* juveniles and a high release of *U. linza* sporelings (up to ~ 90% at 20 Pa shear stress). Moreover, they inhibit the settlement of *B. amphitrite* better than the PDMS control [77].
$(CF_2)_6F$ ... $O-CH_3$ ... $F(F_2C)_6$	Amphiphilic PEG-perfluorohexyl acrylate diblock copolymers resist adsorption of human serum albumin and calf serum much better than do pentablock copolymers containing polysiloxane block. Blending of the pentablock copolymer with a PDMS matrix results in increased protein adsorption. Differently, PDMS-based films with 4–8 wt% pentablock copolymer show high removal percentages (up to ~ 70% at 52 Pa shear stress) of sporelings of *U. linza* [78].
$C_4H_9-Si(O-Si)_m$ ... $C_4H_9-Si(O-Si)_m$ ... $(CF_2)_6F$	Films containing 7 wt% fluorine-free amphiphilic copolymer show a percentage removal of *U. linza* sporelings of ~ 90% at 52 Pa shear stress and perform better than the structurally similar fluorinated amphiphilic terpolymer [79].
$C_4H_9-Si(O-Si)_z$ ... $(CF_2)_8F$	Films containing 1–4 wt% copolymer with lower amount of perfluorooctyl-PEGylated counits exhibit very good FR properties against *U. linza* sporelings (up to ~ 80% removal at 52 Pa shear stress) and excellent AF capability against the settlement of *B. amphitrite* cyprids. Moreover, the detachment of adult barnacles is easier than for the PDMS control. Raft panels immersed in static conditions show significantly reduced levels of hard fouling compared to the PDMS control, their performance being equivalent to the commercial paint Intersleek 700 [76,80,81].
$C_4H_9-Si(O-Si)_z$ ... $(CF_2)_8F$	Removal percentage (up to ~ 75% at 13.6 Pa shear stress) of *U. linza* sporelings is higher for all the copolymer containing films compared to the control PDMS. Moreover, increasing the amount of copolymer from 1 to 10 wt% in the formulation results in an increase in sporeling release [82].

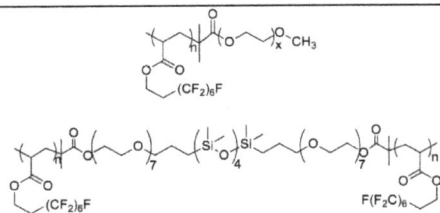

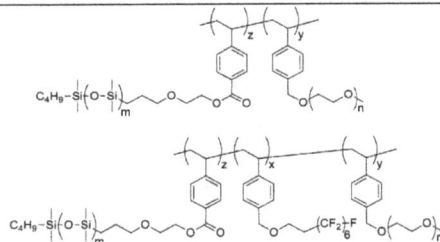

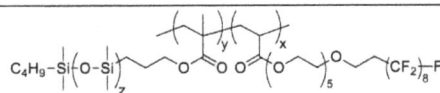

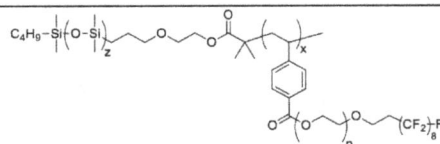

photopolymerized with a polysiloxane carrying methacrylic side groups (Figure 6.8). The obtained crosslinked films represented an elastomeric nature (storage modulus <5 MPa) due to the high content (>90 wt%) of polysiloxane macromonomer in the formulation. By changing the relative ratio of PEG additives and fluoroalkyl additives, films with varied degrees of amphiphilicity were obtained, and their FR potential was tested against the serpulid *Ficopomatus enigmaticus,* a new model organism, and diatom *N. salinicola* [83]. Condensation-cured amphiphilic PDMS coatings were also prepared by using PEG- and fluoroalkyl-trialkoxysilanes as modifying agents [84] (Figure 6.8). Similarly, blends of bissilanol-terminated PDMS, 2-[methoxy(polyethyleneoxy)propyl]trimethoxysilane (TMS-PEG), bissilanol-terminated polytrifluoropropylmethylsiloxane (CF$_3$-PDMS) macromonomers and silanized silica particles were investigated using combinatorial/high-throughput experiments [85]. The bacteria *Cellulophaga lytica* and *Halomonas pacifica* were almost

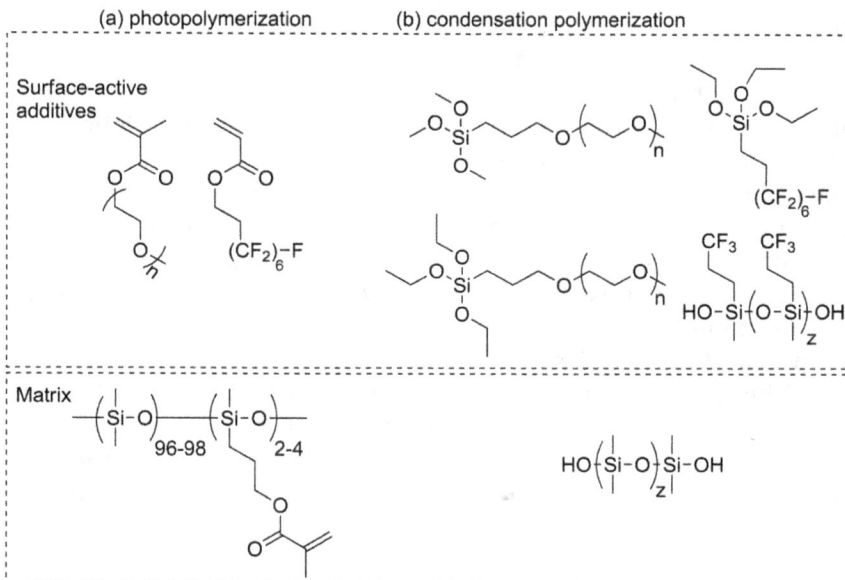

Figure 6.8 Chemical structures of reactive, surface-active additives and PDMS matrices used for the construction of amphiphilic PDMS-based networks via photopolymerization [83] (a) and condensation polymerization [84,85] (b).

completely removed from films containing the highest amounts of both TMS-PEG and CF$_3$-PDMS, which performed much better than the networks with only one of the surface-active agents. The same result was also observed for the release of reattached adult barnacles. This was attributed to a synergistic effect between CF$_3$-PDMS and TMS-PEG. In particular, the latter was dragged to the polymer surface by CF$_3$-PDMS, the lowest surface energy component in the formulation.

Fluorine-free condensation cured amphiphilic PDMS networks composed of dihydroxyl-terminated PDMS and non-conventional branched [86] or linear [87] poly(ethylene oxide) (PEO)-silane amphiphiles containing oligodimethyl siloxane (ODMS) tethers and crosslinkable triethoxysilane groups with varied lengths of ODMS (m = 0, 4, 13) and PEO (n = 3, 8, 16) were prepared (Figure 6.9). Both hydrophobicity and high flexibility of the siloxane spacer promoted the migration of the PEO segments to the polymer–water interface as confirmed by extensive atomic force microscopy [88] and contact angle analyses [87,89]. The extent of surface reconstruction upon immersion in water was found to depend on both the siloxane (m) [87] and PEO (n) chain lengths [90], with the PEO-silane amphiphile with m = 13 and n = 8 generally forming the most hydrophilic and protein resistant surfaces. In contrast, conventional PEO-silane amphiphiles containing short alkyl spacers had a more limited tendency to segregate to the outermost surface layers after immersion in water [91-94]. Besides protein resistant properties, silicone coatings containing non-conventional PEO-silane amphiphiles with the longest siloxane tether displayed improved ability to inhibit the settlement of the marine bacterium *Bacillus* sp.416 and the diatom *Cylindrotheca closterium*, as well as a community of the two [95]. These laboratory assays were repeated on coatings with varied content (1-5 wt%) of the best performing PEO-silane amphiphile (m = 13) and revealed the enhanced antifouling properties of the coatings with 5 and 10 wt% modifying agent. The results were also confirmed by on-site evaluation of microfouling on test panels immersed in the Atlantic Ocean for extended periods of time [96]. In a recent study, structurally similar reactive and non-reactive PEO-silanes, with the same length of PEO (n = 8) and two different lengths of siloxane (m = 13, 30) were compared. Results indicate that protein adsorption was generally low. Interestingly, copolymer leaching and water uptake were similar for both the reactive and non-reactive additives, indicating that especially for the longer siloxane tether, the need for alkoxysilane functionalities was not necessary given the enhanced compatibility with the polymer matrix, which further prevented copolymer leaching [70].

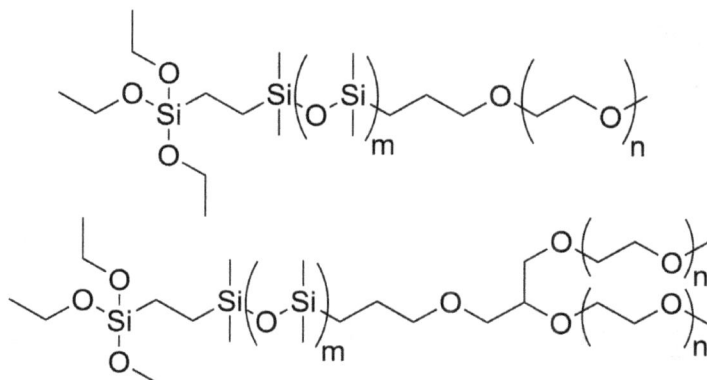

Figure 6.9 Chemical structures of reactive and surface-active PEO-silane amphiphiles [86,87].

6.5.3 Zwitterionization of Polysiloxane Surfaces

Zwitterions are electrically neutral molecules that possess equal amounts of negatively and positively charged groups. Recently, zwitterionic polymers containing phosphorylcholine, sulfobetaine or carboxybetaine groups were considered as antifouling systems due to their ability to strongly bind water molecules via electrostatically induced hydration [97-99], resulting in a hydration layer stronger than that of PEG [100-102], which is thermodynamically unfavorable for the adsorption or adhesion of fouling species. Compared to PEG, zwitterions are more stable to oxidative degradation in the marine environment and, thus, appear to be more suitable for long-term use in such conditions [103-105]. Zwitterions were reported to reduce the non-specific proteins adsorption, resist cell adhesion [106] and be effective against marine organisms. For example, poly(sulfobetaine methacrylate) brushes onto glass surfaces showed AF/FR performance against zoospores of the green alga *U. linza* and the diatom *N. incerta* [107], along with resisting the settlement of *B. amphitrite* [108].

　　The modification of PDMS with zwitterions has also shown promise to enhance fouling resistance properties, thanks to the antifouling active mode of the polyzwitterions. Both "grafting-from" [109] and "grafting-to" [110] methods and post-modification of functional prepolymers were reported for the covalent anchorage of zwitterionic chains to PDMS [111] (Figure 6.10).

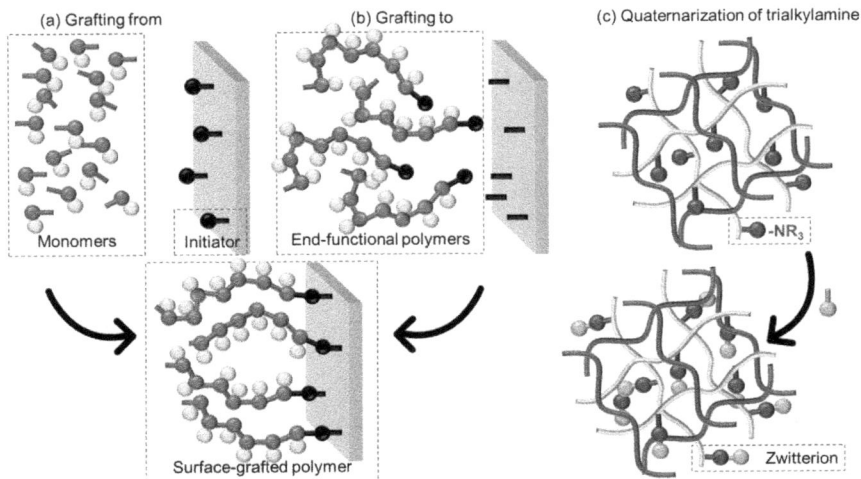

Figure 6.10 Schematic illustration of three different approaches for the zwitterionization of PDMS film surfaces: grafting from (a), grafting to (b) and post-modification of amphiphilic trialkylamine containing networks (c).

In one example, a bromine-terminated alkylthiol molecule was immobilized on the surface of poly(vinylmethylsiloxane) (PVMS) and used for the grafting of poly(sulfobetaine methacrylate) (poly(SBMA)) via activators regenerated by electron transfer atom transfer radical polymerization (ARGET-ATRP) (Figure 6.11) [112]. The polyzwitterion-modified PDMS surfaces were proven to inhibit the settlement of bacteria and barnacle cyprids over short-term exposure, along with promoting the removal of the biofilm eventually formed after long-term exposure.

In a completely different approach, Wang *et al.* [113] prepared a series of amphiphilic co-networks by crosslinking linear PDMS-poly(*N*,*N*-dimethylaminoethyl methacrylate) (PDMAEMA) pentablock copolymers containing pendant alkene functionalities via thiol–ene click chemistry. The spontaneously surface segregated PDMAEMA copolymer was post-reacted with 3-bromopropionic acid for the formation of carboxybetaine functionalities at the surface (Figure 6.12). The AF efficiency of the coatings against *Phaeodactylum tricornutum*, a widespread diatom, was attributed to a combination of an antifouling active mode due to the zwitterionic moieties and a passive mode due to the presence of a nanoscopically complex surface.

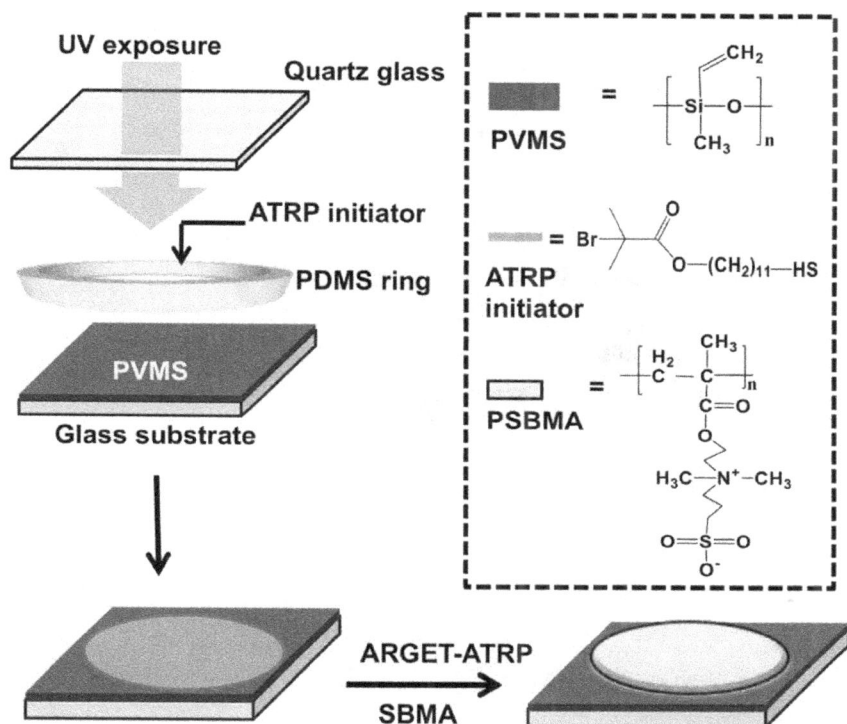

Figure 6.11 Grafting-from of poly(SBMA) on a PVMS elastomer film cured on a glass substrate and functionalized with an ATRP initiator contained in a PDMS ring. The ATRP-initiator from which the ARGET-ATRP of SBMA follows is immobilized on the top of the PVMS film by exposure to UV light. Reproduced from Reference 112 with permission from American Chemical Society.

Charged AF/FR amphiphilic PDMS-based coatings have also been reported in recent studies [114,115]. In one example, co-acrylate surface grafts of different molecular weights, composed of acrylic acid (AA), acrylamide (AAm) and methyl acrylate (MA), were tested against zoospores of the diatom *N. incerta* and the green alga *U. linza*. The coatings displayed better AF/FR properties against zoospores and diatoms, as compared to the non-modified PDMS. However, co-polymer functionalization of siloxane film modified with the Sharklet™ microtopography and variation of graft molar masses did not result in a significant improvement in biological performance of the system [115].

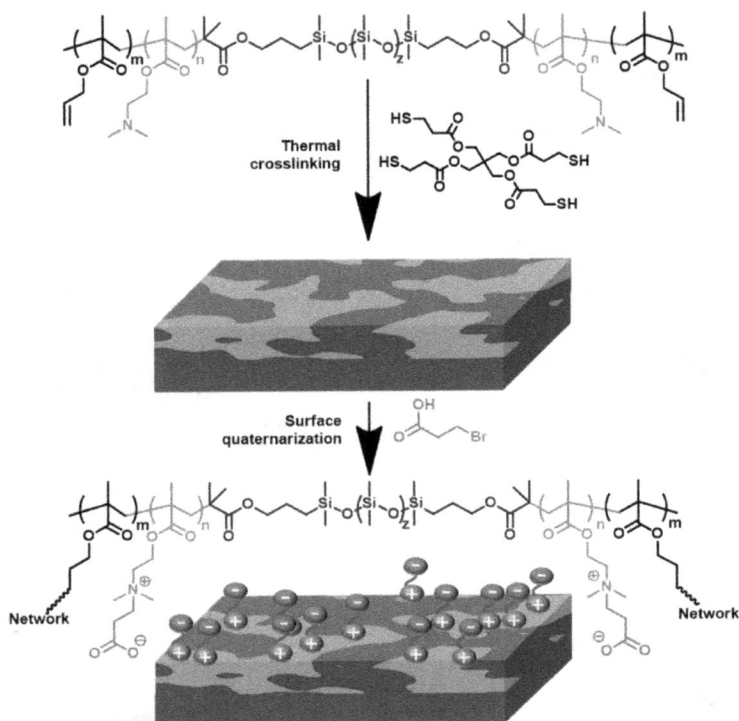

Figure 6.12 Schematic illustration of the zwitterionization of a crosslinked network derived from a PDMAEMA-based pentablock copolymer. Adapted from Reference 113 with permission from American Chemical Society.

6.5.4 Hydrophilization of Self-stratified Siloxane-Polyurethane Coatings

As already discussed above, self-stratified polysiloxane-polyurethane (SiPU) coatings combine in the same systems the low surface energy, low elastic modulus and related FR properties of the siloxane layer at the surface with the bulk mechanical performance and excellent adhesion of polyurethane to the substrate. In an early attempt to impart amphiphilicity to the otherwise hydrophobic nature of SiPU films, a PDMS functionalized with pendant hydrophilic carboxylic acid groups was incorporated into a polyurethane or polyurethane-urea matrix (Figure 6.13) [116]. As a result of the self-stratification process of the PDMS component, the carboxylic acid moieties were also dragged to the surface. The obtained amphiphilic coatings showed improved FR efficacy for microfoulers, such as the diatom *N. incerta,* whereas the removal of marine bacterial biofilm and *U. linza* sporelings was found to be similar to or lower than that of the non-fuctionalized SiPU control.

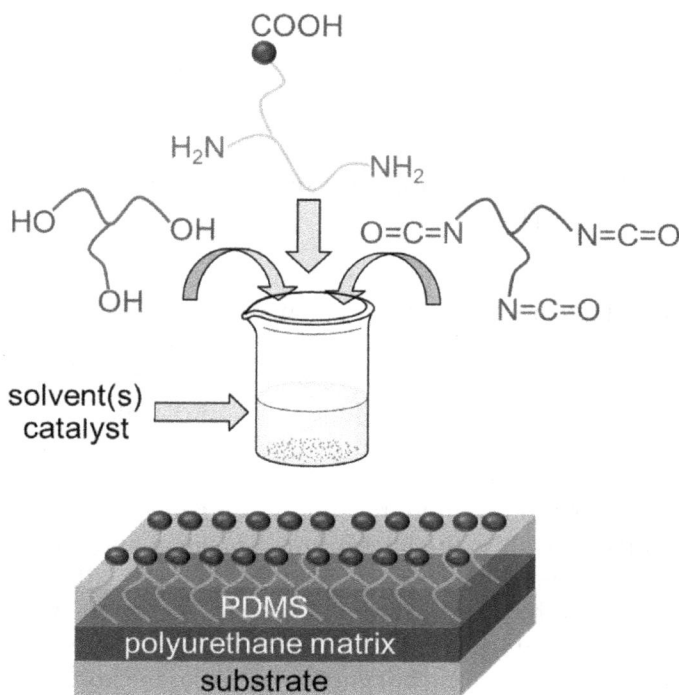

Figure 6.13 Schematic of preparation of a self-stratified polysiloxane-polyurethane (SiPU) coating with pendant carboxylic acid groups. Redrawn from Reference 116.

In another strategy, aminopropyl-terminated siloxanes with PEG grafts provided amphiphilic SiPU coatings with enhanced release of microalgae, but reduced performance against macrofouling [117]. Better biological performances were achieved for amphiphilic SiPU coatings prepared using a series of isophorone diisocyanate trimer-based prepolymers modified with PDMS and PEG chains (Figure 6.14) [118]. The obtained coatings displayed similar or superior FR performance against the macroalga *U. linza*, the diatom *N. incerta* and the bacterium *C. lytica*, as compared to different commercially-available FR coatings.

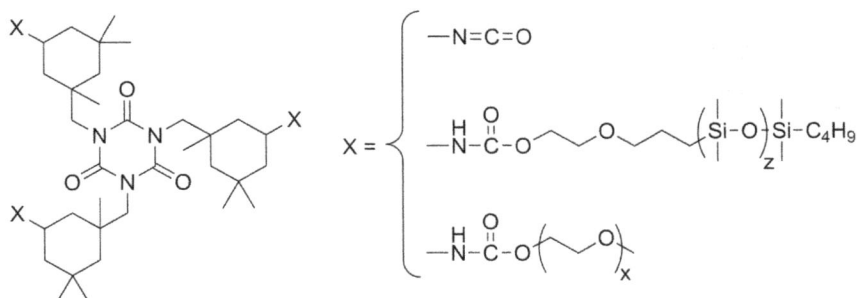

Figure 6.14 Chemical structure of polyisocyanate prepolymers modified with PEG and PDMS [118].

An increase in the hydrophilic nature of SiPU coatings was also achieved by incorporation of a pre-synthesized ABA triblock copolymer, poly(SBMA)-*b*-PDMS-*b*-poly(SBMA), composed of a PDMS central block and poly(sulfobetaine methacrylate) side blocks with secondary amine groups at the junction points capable of reacting with a polyisocyanate to covalently anchor the block copolymer to the SiPU coating derived therefrom (Figure 6.15) [110]. Coating tests against a suite of model marine organisms showed very good FR performance against the bacterium *H. pacifica* and the diatom *N. incerta*. In another approach, zwitterion-tethered SiPU networks were synthesized by condensation of a di/triisocyanate, siloxane diol and carboxybetaine-based diol with a *t*-butyl protecting group. After the hydrolysis of carboxybetaine ester, zwitterionic groups were introduced into the SiPU networks, and the coatings were proven to exhibit excellent resistance to bacterial adhesion and non-specific protein adsorption [119].

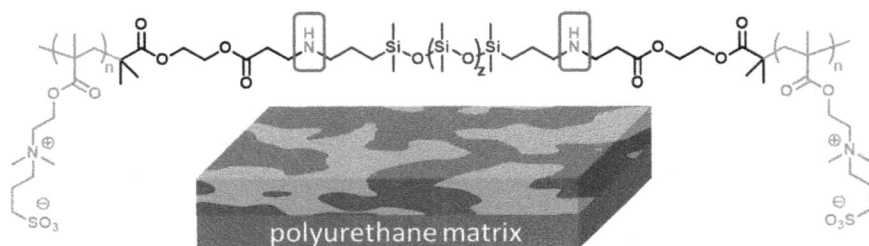

Figure 6.15 Schematic illustration of the zwitterionic SiPU coatings obtained by incorporation of poly(SBMA)-*b*-PDMS-*b*-poly(SBMA) triblock copolymer into a polyurethane coating formulation [110].

6.5.5 Modification of Polysiloxane with Biomolecules and Biomimetic Molecules

Peptides are biological oligomers and polymers composed of aminoacid monomer residues linked by peptide (amide) bonds. A wide array of natural and non-natural peptides can be exploited as amphiphilic building blocks for applied self-assembled nanoarchitectures [120]. On the other hand, peptoids are non-biologically occurring peptide-mimic oligomers composed of *N*-substituted building blocks in which the alkyl group is linked to the nitrogen atom instead of the *α*-carbon, as in the corresponding amino acid (Figure 6.16). This structural difference between the two constitutional isomers makes peptoids more easily thermal and solution processable owing to the lack of intermolecular hydrogen bonds. Self-assembled monolayers (SAMs) of both peptides and peptoids were reported to significantly limit the adsorption of non-specific proteins [121-125]. Moreover, peptide and peptoid chemistries provide a powerful tool to study how variations in composition and sequence distribution at the molecular level impact on the AF/FR performance of the coatings and can be used to optimize coating formulations based on components commonly in use.

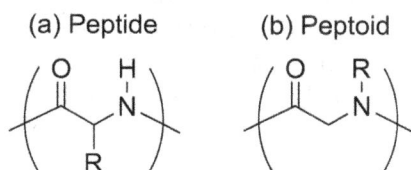

(a) Peptide   (b) Peptoid

Figure 6.16 Chemical structure of repeating unit of peptides (a) and peptoids (b).

In a recent study [126], polystyrene-*b*-poly(dimethylsiloxane-*co*-vinylmethylsiloxane)-*b*-polystyrene   (PS-*b*-P(DMS-*co*-VMS)-*b*-PS) and   polystyrene-*b*-poly(ethyleneoxide-*co*-allylglycidyl   ether)-*b*-polystyrene (PS-*b*-P(EO-*co*-AGE)-*b*-PS) triblock copolymers were synthesized and functionalized with sequence-controlled oligopeptide and oligopeptoid side chains. These were used as amphiphilic surface-active modifiers of a SEBS elastomer film with the main objective to compare the effects of hydrophilicity, hydrogen bonding and monomer sequence on the AF/FR performance of the two systems. Biological assays showed that peptoid side chains favored the removal of *U. linza* sporelings from the PDMS-based copolymer. The lower adhesion strength of sporelings to these surfaces was attributed to the lack of a hydrogen bond donor in the peptoid backbone. On the other hand, PEO-based coatings showed lower initial attachment and adhesion strength of the diatom *N. incerta* with respect to the PDMS counterpart. This work appears to be the continuation of previous work from Ober and coworkers where peptides and peptoids on PDMS-based [127] and PEO-based [128-130] diblock copolymers, respectively, were utilized as amphiphilic sequence-defined oligomers to modify the biological properties of a SEBS elastomeric material. More recently, the triblock copolymers PS-*b*-P(DMS-*co*-VMS)-*b*-PS and PS-*b*-P(EO-*co*-AGE)-*b*-PS with a hydrophobic and hydrophilic central block, respectively, were alternatively functionalized by thiol-ene click chemistry with fluorinated or non-fluorinated synthetic oligopeptides. The modified PDMS-based films performed well against the *U. linza* sporelings, and the PEO-based films exhibited effective performance against both *U. linza* spores and diatoms of *N. incerta*, with the fluorinated oligopeptide-modified PEO based films displaying overall superior antifouling and fouling-release performance than the PDMS standard for both marine organisms. [131].

An example of an aminoacid-modified-PDMS was recently described consisting of a cysteine-grafted poly(DMS-*b*-VMS) block copolymer, which exhibited a high resistance to the adsorption of bovine serum albumin (BSA), even though the AF/FR performance against marine organisms was not tested (Figure 6.17) [132].

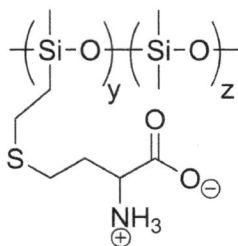

Figure 6.17 Chemical structure of the cysteine-grafted poly(DMS-*b*-VMS) block copolymer [132].

Another attractive approach is to use enzymes as antifouling active biomacromolecules by taking advantage of their biodegradable nature and ability to degrade a wide range of proteins, representing actually an environmentally-friendly alternative to traditional bio-cides for marine coatings [133-135]. The antifouling mechanisms commonly reported for enzymes include degradation of the adhe-sives secreted by marine organisms [136], hydrolysis of cell tissues [133] and production of hydrogen peroxide [133], which has toxic effects on some fouling species. Kim *et al.* [137] reported in an early study that hydrolytic enzymes, namely pronase and α-chymotrypsin, immobilized onto a condensation cured PDMS coating either by sol-gel entrapment or covalent attachment, effectively inhibited the ad-sorption of BSA, chosen as a model protein (Figure 6.18).

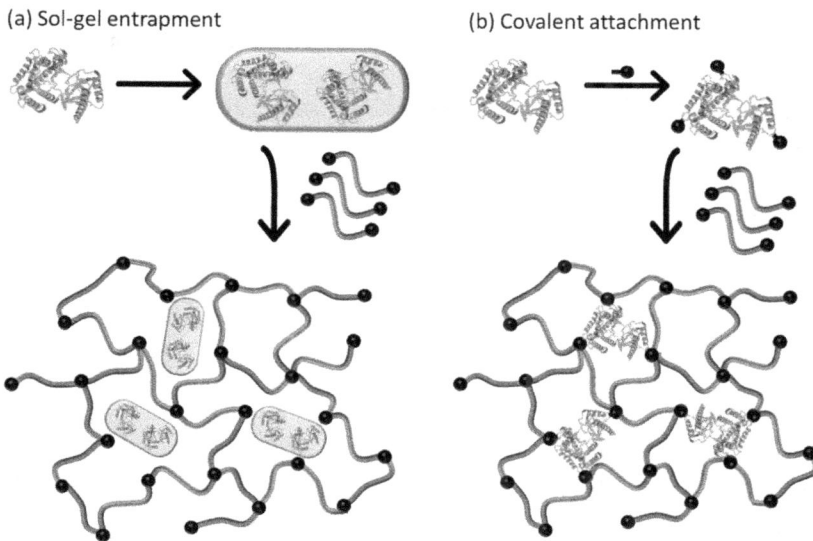

(a) Sol-gel entrapment (b) Covalent attachment

Figure 6.18 Immobilization of enzyme onto PDMS polymers via sol-gel entrapment (a) and covalent attachment (b) processes. Redrawn from Reference 137.

Surface-functionalized microstructured PDMS substrates were modified with an enzyme grafted hydrogel network composed of methacrylic acid and PEG dimethacrylate in order to maintain the surface hydrophilicity and self-cleaning ability against protein ad-sorption [138]. Finally, Olsen and Yebra [139] patented a marine coating formulation consisting of a polysiloxane-based binder sys-tem comprising one or more polysiloxane components modified with hydrophilic oligomer/polymer moieties, and one or more pro-teolytic enzymes [139].

6.6 Other Strategies to Improve the AF/FR Performance of PDMS-based Coatings

6.6.1 Self-polishing PDMS-based Coatings

Recently, novel biocide-free antifouling coatings consisting of a hydrolyzable poly(trialkylsilyl methacrylate) component and a hydrophobic/low surface energy PDMS component were synthesized to produce hybrid self-polishing (SP)/fouling release (FR) coatings. In particular, Bressy and coworkers prepared copolymers with diverse well-defined architectures (random, block, graft) by reversible addition-fragmentation chain-transfer polymerization (RAFT), starting from SP and FR comonomers with different molecular structures (Figure 6.19). The antifouling properties of these copolymers were described for both laboratory assays and static field immersion trials in the Toulon Bay [140-142]. This strategy can be further exploited to prepare amphiphilic SP/FR copolymers to be dispersed in a crosslinkable PDMS matrix as either non-reactive or reactive surface-active additives.

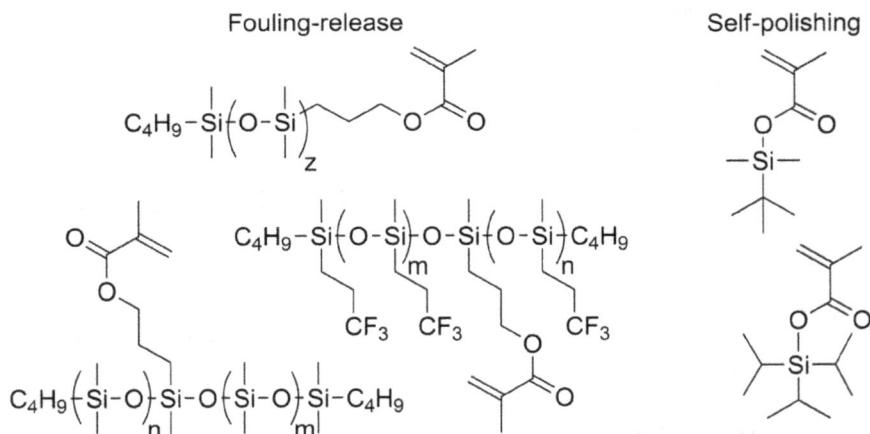

Figure 6.19 Chemical structures of FR and SP comonomers [140-142].

6.6.2 PDMS Coatings Containing Tethered Biocides

Biocides are generally very active against a broad spectrum of marine species when released from a coating. In order to overcome their toxicity drawbacks and employ them in a more environmental-ly-friendly manner, one strategy is to covalently bind the biocide to the polymer matrix [143,144]. PDMS is one of the most common ma-terials used for functionalization. By this approach, it is possible to combine the FR performance of the polymer matrix with the AF properties of the biocide, thus, simultaneously preventing the re-lease and reducing the bioaccumulation of the toxic biocide in the environment.

Polysiloxane networks with tethered quaternary ammonium salt (QAS) were widely studied by Majumdar *et al.* [145-147]. QASs are used as antimicrobial agents and disinfectants. Their high charge density strongly interacts with the negatively charged cell walls of several microorganisms and neutralizes them by contact [148,149]. Siloxane coatings with QAS tethers generally presented heterogene-ous surface morphologies with different levels of nano-roughness depending on the alkyl chain length of the QAS segregated to the polymer surface [150]. QAS were reported to be effective against marine algae and inhibited the adhesion of marine bacteria and dia-toms [150]. In a more recent example, *N*-(2,4,6-trichlorophenyl) ma-leimide (TCM) pendant groups were incorporated in SiPU coatings [151]. TCM is a non-ionic biocide with a low toxicity [152], which, in contrast to QAS, is anticipated to have a low impact on the wettabil-ity of the coatings especially after immersion in seawater. Laborato-ry assays proved that these coatings were effective in reducing the adhesion of both the marine bacterium *Micrococcus luteus* and the diatom *N. incerta*, along with the settlement of the *B. amphitrite* cyprids. The AF properties were also confirmed by field immersion trials in South China Sea.

6.7 Silicone-based Coatings Available in the Market

The development of polysiloxane-based FR technologies has had a direct impact on the evolution of commercial FR coatings, as proved by the market products as well as the patents filed by the leading companies in the field of marine antifouling paints [16,153].

International Paint launched in 2007 a biocide-free FR coating, namely Intersleek 900, consisting of an amphiphilic silox-ane/fluoropolymer, which confers low surface energy, smoothness and slipperiness to the finish coat. Later on, a slime release coating, i.e., Intersleek 1100SR, was marketed in 2013, containing a fluoro-polymer additive modified with hydrophilic moieties. More recently,

Intersleek 1000 was launched in 2016, in which a bio-renewable raw material (lanolin) is combined with the silicone technology. The use of fluorinated [154] and fluorinated oxyalkylene-containing [155] polymers or oligomers as surface-active additives has been patented. The use of lanolin and lanolin derivatives was also patented [156].

The first silicone product introduced in the market in 2000 by Hempel A/S was Hempasil 77500, a biocide-free hydrophobic low surface energy siloxane coating. This was replaced in 2008 by Hempasil X3, which combines the hydrogel aspect with the silicone FR technology through the dispersion of a hydrophilic modified silicone oil in a condensation-cured matrix. The amphiphilic surface-active additive maintains a more hydrophilic surface, which guarantees a better control over diatom slime. Polyethyleneglycol, polypropyleneglycol, poly(acrylic acid) and poly(vinyl pyrrolidone) were reported as the hydrophilic components of the modified silicone. In Hempaguard X7 (launched in 2013), the previous hydrogel silicone technology was combined with the release of small amounts (<10% compared to conventional SPC) of booster biocides, such as copper pyrithione and seanine [157,158].

Chugoku Marine Paints (CMP) launched in the early 2000s a biocide-free coating, Bioclean, containing a "hydrophobic oil" [159]. The advanced version of the product, Bioclean Plus (released in 2014), is still based on a silicone elastomer technology, but with the incorporation of a small amount of antifouling active agent (i.e., biocide). The use of phenyl or polyether-modified polysiloxanes with a hydrolyzable end-group as silicon oils and pyrithione metal salts as antifouling agents was also reported [160].

SeaLion is the FR product launched in the market by Jotun in 2005, based on hydrophobic low surface energy siloxane technology. SeaLion Repulse, commercialized in 2010, is a condensation curing coating with an improved release of biofouling, thanks to a film surface engineered with "nanoscale springs" [16]. Jotun also reported the use of (meth)acrylic copolymers carrying perfluoroalkyl and siloxane grafts as non-reactive or reactive (through the incorporation of alkoxysilane groups) surface-active agents in the PDMS-based formulation [51].

6.8 Concluding Remarks

The use of antifouling paints is of prime importance to combat marine biofouling and face the associated economical costs as well as environmental issues. Although biocide-based self-polishing coatings still dominate the market, their use is becoming increasingly more regulated and non-toxic FR PDMS-based coatings actually represent the highest promising alternative. To enhance their resistance to microfoulers, especially slimes, several strategies have been proposed. All such approaches share the underpinning concept to increase the hydrophilic nature of the otherwise hydrophobic PDMS surface by the incorporation in the formulation of active AF moieties, including PEG segments, zwitterions, biomacromolecules (peptides and enzymes) and biomimetic molecules (peptoids). The hydrophilic/hydrophobic (i.e. amphiphilic) system derived therefrom displays also a manifold character with a mixed surface chemical composition, morphology and topography at the micro- to nano-scale level that drastically affects in a passive mode the settlement and adhesion of foulers.

The potential of chemically hydrophilized FR PDMS-based systems as eco-sustainable antifouling coatings has been widely demonstrated by the AF/FR laboratory assays against different micro- and macro-foulers, used as representative marine organisms. However, not all of the amphiphilic silicone-based approaches presented in this chapter are suitable for implementation on an industrial large-scale production (e.g. shipping industry), as these do not meet specific requirements, including low cost of production, easy application, long term lifetime in service, etc. Overall, the addition of amphiphilic surface-active modifiers is the most followed technology for the development of PDMS-based FR commercial coatings, which have recently experienced a significant expansion in the market.

References

1. Schultz, M. P. (2007) Effects of coating roughness and biofouling on ship resistance and powering. *Biofouling*, **23**, 331-341.
2. Schultz, M. P., Bendick, J. A., Holm, E. R., and Hertel, W. M. (2011) Economic impact of biofouling on a naval surface ship. *Biofouling*, **27**, 87-98.
3. Poloczanska, E. S., and Butler, A. J. (2010) Biofouling and climate change. In: *Biofouling*, Durr, S., and Thomason, J. C. (eds), Wiley-Blackwell, USA, pp. 333-347.
4. Piola, R. F., Dafforn, K. A., and Johnston, E. L. (2009) The influence

of antifouling practices on marine invasions. *Biofouling*, **25**, 633-644.

5. McCollin, T., and Brown, L. (2014) Native and non native marine biofouling species present on commercial vessels using Scottish dry docks and harbours. *Management of Biological Invasions*, **5**, 85-96.

6. Yebra, D. M., Kiil, S., and Dam-Johansen, K. (2004) Antifouling technology-past, present and future steps towards efficient and environmentally friendly antifouling coatings. *Progress in Organic Coatings*, **50**, 75-104.

7. Okoro, H. K., Fatoki, O. S., and Adekola, F. A. (2011) Sources, environmental levels and toxicity of organotin in marine environment- A review. *Asian Journal of Chemistry*, **23**, 473-482.

8. Pereira, M., and Ankjaergaard, C. (2009) Legislation affecting antifouling products. In: *Advances in Marine Antifouling Coatings and Technologies,* Hellio, C., and Yebra, D. (eds.), Woodhead Publishing Limited, UK, pp. 240-259.

9. Ciriminna, R., Bright, F. V., and Pagliaro, M. (2015) Ecofriendly antifouling marine coatings. *ACS Sustainable Chemistry & Engineering*, **3**, 559-565.

10. Carson, R. T., Damon, M., Johnson, L. T., and Gonzalez, J. A. (2009) Conceptual issues in designing a policy to phase out metal-based antifouling paints on recreational boats in San Diego Bay. *Journal of Environmental Management*, **90**, 2460-2468.

11. Banerjee, I., Pangule, R. C., and Kane, R. S. (2011) Antifouling coatings: recent developments in the design of surfaces that prevent fouling by proteins, bacteria, and marine organisms. *Advanced Materials*, **23**, 690-718.

12. Krishnan, S., Weinman, C. J., and Ober, C. K. (2008) Advances in polymers for anti-biofouling surfaces. *Journal of Materials Chemistry*, **18**, 3405-3413.

13. Nurioglu, A. G., Esteves, A. C. C., and de With, G. (2015) Non-toxic, non-biocide-release antifouling coatings based on molecular structure design for marine applications. *Journal of Materials Chemistry B*, **3**, 6547-6570.

14. Callow, J. A., and Callow, M. E. (2011) Trends in the development of environmentally friendly fouling-resistant marine coatings. *Nature Communications*, **2**, 244.

15. Omae, I. (2003) General Aspects of Tin-Free Antifouling Paints. *Chemical Reviews*, **103**, 3431-3448.

16. Lejars, M., Margaillan, A., and Bressy, C. (2012) Fouling release coatings: A nontoxic alternative to biocidal antifouling coatings. *Chemical Reviews*, **112**, 4347-4390.

17. Baier, R. E. (2006) Surface behaviour of biomaterials: the theta surface for biocompatibility. *Journal of Materials Science: Materials*

in Medicine, **17**, 1057-1062.

18. Brady, R. F., and Singer, I. L. (2000) Mechanical factors favoring release from fouling release coatings. *Biofouling*, **15**, 73-81.

19. Kohl, J. G., and Singer, I. L. (1999) Pull-off behavior of epoxy bonded to silicone duplex coatings. *Progress in Organic Coatings*, **36**, 15-20.

20. Singer, I. L., Kohl, J. G., and Patterson, M. (2000) Mechanical aspects of silicone coatings for hard foulant control. *Biofouling*, **16**, 301-309.

21. Chung, J. Y., and Chaudhury, M. K. (2005) Soft and hard adhesion. *The Journal of Adhesion*, **81**, 1119-1145.

22. Kendall, K. (1971) The adhesion and surface energy of elastic solids. *Journal of Physics D: Applied Physics*, **4**, 1186-1195.

23. Clare, A. S., and Hoeg, J. T. (2008) *Balanus amphitrite* or *Amphibalanus amphitrite*? A note on barnacle nomenclature. *Biofouling*, **24**, 55-57.

24. Chaudhury, M. K., Finlay, J. A., Chung, J. Y., Callow, M. E., and Callow, J. A. (2005) The influence of elastic modulus and thickness on the release of the soft-fouling green alga *Ulva linza* (syn. *Enteromorpha linza*) from poly(dimethylsiloxane) (PDMS) model networks. *Biofouling*, **21**, 41-48.

25. Wendt, D., Kowalke, G., Kim, J., and Singer, I. (2006) Factors that influence elastomeric coating performance: the effect of coating thickness on basal plate morphology, growth and critical removal stress of the barnacle *Balanus amphitrite*. *Biofouling*, **22**, 1-9.

26. Townsin, R. L., and Anderson, C. D. (2009) Fouling control coatings using low surface energy, foul release technology. In: *Advances in Marine Antifouling Coatings and Technologies*, Hellio, C., and Yebra, D. (eds.), CRC Press, USA, pp. 693-708.

27. Eduok, U., Faye, O., and Szpunar, J. (2017) Recent developments and applications of protective silicone coatings: A review of PDMS functional materials. *Progress in Organic Coatings*, **111**, 124-163.

28. Chen, C., Wang, J., and Chen, Z. (2004) Surface restructuring behavior of various types of poly(dimethylsiloxane) in water detected by SFG. *Langmuir*, **20**, 10186-10193.

29. Thomas, J., Choi, S.-B., Fjeldheim, R., and Boudjouk, P. (2004) Silicones containing pendant biocides for antifouling coatings. *Biofouling*, **20**, 227-236.

30. Pike, J. K., Ho, T., and Wynne, K. J. (1996) Water-induced surface rearrangements of poly (dimethylsiloxaneurea-urethane) segmented block copolymers. *Chemistry of Materials*, **8**, 856-860.

31. Pretti, C., Oliva, M., Mennillo, E., Barbaglia, M., Funel, M., Yasani, B. R., Martinelli, E., and Galli, G. (2013) An ecotoxicological study on tin- and bismuth-catalysed PDMS based coatings containing a surface-active polymer. *Ecotoxicology and Environmental Safety*, **98**,

250-256.

32. Milne, A. (1977) Anti-Fouling Marine Compositions, patent US4025693A.

33. Edwards, D. P., Nevell, T. G., Plunkett, B. A., and Ochiltree, B. C. (1994) Resistance to marine fouling of elastomeric coatings of some poly (dimethylsiloxanes) and poly(dimethyldiphenyl-siloxanes). *International Biodeterioration & Biodegradation*, **34**, 349-359.

34. Truby, K., Wood, C. D., Stein, J., Cella, J., Carpenter, J., Kavanagh, C., Swain, G., Wiebe, D., Lapota, D., Meyer, A., Holm, E., Wendt, D., Smith, C., and Montemarano, J. (2000) Evaluation of the performance enhancement of the silicone biofouling-release coatings by oil incorporation. *Biofouling*, **15**, 141-150.

35. Stein, J., Truby, K., Wood, C. D., Stein, J., Gardner, M., Swain, G., Kavanagh, C., Kovach, B., Schultz, M., Wiebe, D., Holm, E., Montemarano, J., Wendt, D., Smith, C., and Meyer, A. (2003) Silicone foul release coatings: effect of the interaction of oil and coating functionalities on the magnitude of macrofouling attachment strengths. *Biofouling*, **19**, 71-82.

36. Galhenage, T. P., Hoffman, D., Silbert, S. D., Stafslien, S. J., Daniels, J., Miljkovic, T., Finlay, J. A., Franco, S. C., Clare, A. S., Nedved, B. T., Hadfield, M. G., Wendt, D. E., Waltz, G., Brewer, L., Teo, S. L.-M., Lim, C.-S., and Webster, D. C. (2016) Fouling-release performance of silicone oil modified siloxane-polyurethane coatings. *ACS Applied Materials and Interfaces*, **8**, 29025-29036.

37. Holland, R., Dugdale, T. M., Wetherbee, R., Brennan, A. B., Finlay, J. A., Callow, J. A., and Callow, M. E. (2004) Adhesion and motility of fouling diatoms on a silicone elastomer. *Biofouling*, **20**, 323-329.

38. Molino, P. J., Campbell, E., and Wetherbee, R. (2009) Development of the initial diatom microfouling layer on antifouling and fouling-release surfaces in temperate and tropical Australia. *Biofouling*, **25**, 685-693.

39. Molino, P. J., Childs, S., Eason Hubbard, M. R., Carey, J. M., Burgman, M. A., and Wetherbee, R. (2009). Development of the primary bacterial microfouling layer on antifouling and fouling release coatings in temperate and tropical environments in Eastern Australia. *Biofouling*, **25**, 149-162.

40. Dalton, H. M., Stein, J., and March, P. E. (2000) A biological assay for detection of heterogeneities in the surface hydrophobicity of polymer coatings exposed to the marine environment. *Biofouling*, **15**, 83-94.

41. Stein, J., Truby, K., Wood, C. D., Takemori, M., Vallance, M., Swain, G., Kavanagh, C., Kovach, B., Schultz, M., Wiebe, D., Holm, E., Montemarano, J., Wendt, D., Smith, C., and Meyer, A. (2003) Structure-property relationships of silicone biofouling-release coatings: Ef-

fect of silicone network architecture on pseudobarnacle attachment strengths. *Biofouling*, **19**, 87-94.

42. Beigbeder, A., Degee, P., Conlan, S. L., Mutton, R., Clare, A. S., Pettitt, M. E., Callow, M. E., Callow, J. A., and Dubois, P. (2008) Preparation and characterisation of silicone-based coatings filled with carbon nanotubes and natural sepiolite and their application as marine fouling-release coatings. *Biofouling*, **24**, 291-302.

43. Beigbeder, A., Jeusette, M., Mincheva, R., Claes, M., Brocorens, P., Lazzaroni, R., and Dubois, P. (2009) On the effect of carbon nanotubes on the wettability and surface morphology of hydrosilylation-curing silicone coatings. *Journal of Nanostructured Polymers and Nanocomposites*, **5**, 37-43.

44. Ekin, A., Webster, D. C., Daniels, J. W., Stafslien, S. J., Casse, F., Callow, J. A., and Callow, M. E. (2007) Synthesis, formulation, and characterization of siloxane-polyurethane coatings for underwater marine applications using combinatorial high-throughput experimentation. *Journal of Coatings and Technology Research*, **4**, 435-451.

45. Pieper, R. J., Ekin, A., Webster, D. C., Casse, F., Callow, J. A., and Callow, M. E. (2007) Combinatorial approach to study the effect of acrylic polyol composition on the properties of crosslinked siloxane-polyurethane fouling-release coatings. *Journal of Coatings and Technology Research*, **4**, 453-461.

46. Sommer, S., Ekin, A., Webster, D. C., Stafslien, S. J., Daniels, J., VanderWal, L. J., Thompson, S. E. M., Callow, M. E., and Callow, J. A. (2010) A preliminary study on the properties and fouling-release performance of siloxane-polyurethane coatings prepared from poly(dimethylsiloxane) (PDMS) macromers. *Biofouling*, **26**, 961-972.

47. Grunlan, M. A., Lee, N. S., Mansfeld, F., Kus, E., Finlay, J. A., Callow, J. A., Callow, M. E., Weber, W. P. (2006) Minimally adhesive polymer surfaces prepared from star oligosiloxanes and star oligofluorosiloxanes. *Journal of Polymer Science, Part A: Polymer Chemistry*, **44**, 2551-2566.

48. Berglin, M., Wynne, K. J., and Gatenholm, P. (2003) Fouling-release coatings prepared from α,ω-dihydroxypoly(dimethylsiloxane) cross-linked with (heptadecafluoro-1,1,2,2-tetrahydrodecyl) triethoxysilane. *Journal of Colloid Interface Science*, **257**, 383-391.

49. Williams, D. N., Shewring, N. I. E., and Lee, A. J. (2014) Anti-Fouling Compositions with Fluorinated Alkyl- or Alkoxy- Containing Polymer or Oligomer, patent US8771798B2.

50. Marabotti, I., Morelli, A., Orsini, L. M., Martinelli, E., Galli, G., Chiellini, E., Lien, E. M., Pettitt, M. E., Callow, M. E., Callow, J. A., Conlan, S. L., Mutton, R. J., Clare, A. S., Kocijan, A., Donik, C., and Jenko, M. (2009) Fluorinated/siloxane copolymer blends for fouling release:

chemical characterisation and biological evaluation with algae and barnacles. *Biofouling*, **25**, 481-493.

51. Dahling, M., Lien, E., Orsini, L., Galli, G., and Chiellini, E. (2007) Fouling Release Composition, WO2007/102741.

52. Yeh, S.-B., Chen, C.-S., Chen, W.-Y., and Huang, C.-J. (2014) Modification of silicone elastomer with zwitterionic silane for durable antifouling properties, zwitterionic siloxane-polyurethane fouling-release coatings. *Langmuir*, **30**, 11386-11393.

53. Gudipati, C. S., Finlay, J. A., Callow, M. E., Callow, J. A. and Wooley, K. L. (2005) The anti-fouling and fouling-release performance of unique hyperbranched fluoropolymer (HBFP)-poly(ethylene glycol) (PEG) composite coatings evaluated by protein adsorption and the settlement of zoospores of the green fouling alga *Ulva* (syn. *Enteromorpha*). *Langmuir*, **21**, 3044-3053.

54. Krishnan, S., Ayothi, R., Hexemer, A., Finlay, J. A., Sohn, K. E., Perry, R., Ober, C. K., Kramer, E. J., Callow, M. E., Callow, J. A., and Fischer, D. A. (2006) Anti-biofouling properties of comb-like block copolymer with amphiphilic side-chains. *Langmuir*, **22**, 5075-5086.

55. Finlay, J. A., Krishnan, S., Callow, M. E., Callow, J. A., Dong, R., Asgill, N., Wong, K., Kramer, E. J., and Ober, C. K. (2008) Settlement of *Ulva* zoospores on patterned fluorinated and PEGylated monolayer surfaces. *Langmuir*, **24**, 503-510.

56. Galli, G., and Martinelli, E. (2017) Amphiphilic polymer platforms: surface engineering of films for marine antibiofouling. *Macromolecular Rapid Communications*, **38**, 1600704.

57. Narrainen, A. P., Hutchings, L. R., Ansari, I., Thompson, R. L., and Clarke, N. (2007) Multi-End-Functionalized polymers: additives to modify polymer properties at surfaces and interfaces. *Macromolecules*, **40**, 1969-1980.

58. Lee, H., and Archer, L. A. (2002) Functionalizing polymer surfaces by surface migration of copolymer additives: Role of additive molecular weight. *Polymer*, **43**, 2721-2728.

59. Lee, H., and Archer, L. A. (2001) Functionalizing polymer surfaces by field-induced migration of copolymer additives. 1. Role of surface energy gradients. *Macromolecules*, **34**, 4572-4579.

60. Martinelli, E., Fantoni, C., Galli, G., Gallot, B., and Glisenti, A. (2009) Low surface energy properties of smectic fluorinated block copolymer/SEBS blends. *Molecular Crystals and Liquid Crystals*, **500**, 51-62.

61. Inutsuka, M., Yamada, N. L., Ito, K., and Yokoyama, H. (2013) High density polymer brush spontaneously formed by the segregation of amphiphilic diblock copolymers to the polymer/water interface. *ACS Macro Letters*, **2**, 265-268.

62. Krishnan, S., Wang, N., Ober, C. K., Finlay, J. A., Callow, M. E., Callow, J. A., Hexemer, A., Sohn, K. E., Kramer, E. J., and Fischer, D. A. (2006)

Comparison of the fouling release properties of hydrophobic fluorinated and hydrophilic PEGylated block copolymer surfaces: attachment strength of the diatom *Navicula* and the green alga *Ulva*. *Biomacromolecules*, **7**, 1449-1462.

63. Martinelli, E., Agostini, S., Galli, G., Chiellini, E., Glisenti, A., Pettitt, M. E., Callow, M. E., Callow, J. A., Graf, K., and Bartels, F. W. (2008) Nanostructured films of amphiphilic fluorinated block copolymers for fouling release application. *Langmuir*, **24**, 13138-13147.

64. Weinman, C. J., Finlay, J. A., Park, D., Paik, M. Y., Krishnan, S., Sundaram, H. S., Dimitriou, M., Sohn, K. E., Callow, M. E., Callow, J. A., Handlin, D. L., Willis, C. L., Kramer, E. J., and Ober, C. K. (2009) ABC triblock surface active block copolymer with grafted ethoxylated fluoroalkyl amphiphilic side chains for marine antifouling/fouling-release applications. *Langmuir*, **25**, 12266-12274.

65. Dimitriou, M. D., Zhou, Z., Yoo, H.-S., Killops, K. L., Finlay, J. A., Cone, G., Sundaram, H. S., Lynd, N. A., Barteau, K. P., Campos, L. M., Fischer, D. A., Callow, M. E., Callow, J. A., Ober, C. K., Hawker, C. J., and Kramer E. J. (2011) A general approach to controlling the surface composition of poly(ethylene oxide)-based block copolymers for antifouling coatings. *Langmuir*, **27**, 13762-13772.

66. Galli, G., Martinelli, E., Chiellini, E., Ober, C. K., and Glisenti, A. (2005) Low surface energy characteristics of mesophase forming ABC, ACB triblock copolymers with fluorinated B blocks. *Molecular Crystals and Liquid Crystals*, **441**, 211-226.

67. McPherson, T., Kidane, A., Szleifer, I., and Park, K. (1998) Prevention of protein adsorption by tethered poly(ethylene oxide) layers: experiments and single-chain mean-field analysis. *Langmuir*, **14**, 176-186.

68. Li, B., and Ye, Q. (2015) Antifouling surfaces of self-assembled thin layer. In: *Antifouling Surfaces and Materials*, Zhou, F. (ed.), Springer, Germany, pp. 31-54.

69. Wenning, B. M., Martinelli, E., Mieszkin, S., Finlay, J. A., Fischer, D., Callow, J. A., Callow, M. E., Leonardi, A. K., Ober, C. K., and Galli, G. (2017) Model amphiphilic block copolymers with tailored molecular weight and composition in PDMS-based films to limit soft biofouling. *ACS Applied Materials and Interfaces*, **9**, 16505-16516.

70. Rufin, M. A., Ngo, B. K. D., Barry, M. E., Page, V. M., Hawkins, M. L., Stafslien, S. J., and Grunlan, M. A. (2017) Antifouling silicones based on surface-modifying additive amphiphiles. *Green Materials*, **5**, 4-13.

71. Noguer, A. C., Olsen, S. M., Hvilsted, S., and Kiil, S. (2017) Diffusion of surface-active amphiphiles in silicone-based fouling-release coatings. *Progress in Organic Coatings*, **106**, 77-86.

72. Oliva, M., Martinelli, E., Galli, G., and Pretti, C. (2017) PDMS-based films containing surface-active amphiphilic block copolymers to

combat fouling from barnacles *B. amphitrite* and *B. improvisus*. *Polymer*, **108**, 476-482.

73. Martinelli, E., Pelusio, G., Yasani, B. R., Glisenti, A., and Galli, G. (2015) Surface chemistry of amphiphilic polysiloxane/triethyleneglycol-modified poly(pentafluorostyrene) block copolymer films before and after water immersion. *Macromolecular Chemistry and Physics*, **216**, 2086-2094.

74. Martinelli, E., Hill, S. D., Finlay, J. A., Callow, M. E., Callow, J. A., Glisenti, A., and Galli, G. (2016) Amphiphilic modified-styrene copolymer films: antifouling/fouling release properties against the green alga *Ulva linza*. *Progress in Organic Coatings*, **90**, 235-242.

75. Galli, G., Barsi, D., Martinelli, E., Glisenti, A., Finlay, J. A., Callow, M. E., and Callow, J. A. (2016) Copolymer films containing amphiphilic side chains of well-defined fluoroalkyl-segment length with biofouling-release potential. *RSC Advances*, **6**, 67127-67135.

76. Martinelli, E., Sarvothaman, M. K., Galli, G., Pettitt, M. E., Callow, M. E., Callow, J. A., Conlan, S. L., Clare, S. A., Sugiharto, A. B., Davies, C., and Williams, D. (2012) Poly(dimethyl siloxane) (PDMS) network blends of amphiphilic acrylic copolymers with poly(ethylene glycol)-fluoroalkyl side chains for fouling-release coatings. II. Laboratory assays and field immersion trials. *Biofouling*, **28**, 571-582.

77. Martinelli, E., Gunes, D., Wenning, B. M., Ober, C. K., Finlay, J. A., Callow, M. E., Callow, J. A., Di Fino, A., Clare, A. S., and Galli, G. (2016) Effects of surface-active block copolymers with oxyethylene and fluoroalkyl side chains on the antifouling performance of silicone-based film. *Biofouling*, **32**, 81-93.

78. Martinelli, E., Guazzelli, E., Bartoli, C., Gazzarri, M., Chiellini, F., Galli, G., Callow, M. E., Callow, J. A., Finlay, J. A., and Hill, S. (2015) Amphiphilic pentablock copolymers and their blends with PDMS for antibiofouling coatings. *Journal of Polymer Science, Part A: Polymer Chemistry*, **53**, 1213-1225.

79. Yasani, B. R., Martinelli, E., Galli, G., Glisenti, A., Mieszkin, S., Callow, M. E., and Callow, J. A. (2014) A comparison between different fouling-release elastomer coatings containing surface-active polymers. *Biofouling*, **30**, 387-399.

80. Martinelli, E., Sarvothaman, M. K., Alderighi, M., Galli, G., Mielczarski, E., and Mielczarski, J. A. (2012) PDMS network blends of amphiphilic acrylic copolymers with poly(ethylene glycol)-fluoroalkyl side chains for fouling-release coatings. I. Chemistry and stability of the film surface. *Journal of Polymer Science, Part A: Polymer Chemistry*, **50**, 2677-2686.

81. Atlar, M., Unal, B., Unal, U.O., Politis, G., Martinelli, E., Galli, G., Davies, C., Williams, D. (2013) An experimental investigation on the frictional drag characteristics of nanostructured and fluorinated fouling release coatings. *Biofouling*, **29**, 39-52.

82. Martinelli, E., Suffredini, M., Galli, G., Glisenti, A., Pettitt, M. E., Callow, M. E., Callow, J. A., Williams, D., and Lyall, G. (2011) Amphiphilic block copolymer/poly(dimethylsiloxane) (PDMS) blends and nanocomposites for improved fouling-release. *Biofouling*, **27**, 529-541.

83. Martinelli, E., Del Moro, I., Galli, G., Barbaglia, M., Bibbiani, C., Mennillo, E., Oliva, M., Pretti, C., Antonioli, D., and Laus, M. (2015) Photopolymerized network polysiloxane films with dangling hydrophilic/hydrophobic chains for the biofouling release of invasive marine serpulid *Ficopomatus enigmaticus*. *ACS Applied Materials and Interfaces*, **7**, 8293-8301.

84. Martinelli, E., Pretti, C., Oliva, M., Glisenti, A., and Galli, G. (2018) Sol-gel polysiloxane films containing different surface-active trialkoxysilanes for the release of the marine foulant *Ficopomatus enigmaticus*. *Polymer*, **145**, 426-433.

85. Stafslien, S. J., Christianson, D., Daniels, J., VanderWal, L., Chernykha, A., and Chisholm, B. J. (2015) Combinatorial materials research applied to the development of new surface coatings XVI: fouling-release properties of amphiphilic polysiloxane coatings. *Biofouling*, **31**, 135-149.

86. Murthy, R., Bailey, B. M., Valentin-Rodriguez, C., Ivanisevic, A., and Grunlan, M. A. (2010) Amphiphilic silicones prepared from branched PEO-silanes with siloxane tethers. *Journal of Polymer Science, Part A: Polymer Chemistry*, **48**, 4108-4119.

87. Murthy, R., Cox, C. D., Hahn, M. S., and Grunlan, M. A. (2007) Protein-resistant silicones: incorporation of poly(ethylene oxide) via siloxane tethers. *Biomacromolecules*, **8**, 3244-3252.

88. Hawkins, M. L., Rufin, M. A., Raymond, J. E., and Grunlan, M. A. (2014) Direct observation of the nanocomplex surface reorganization of antifouling silicones containing a highly mobile PEO-silane amphiphile. *Journal of Materials Chemistry B*, **2**, 5689-5697.

89. Hawkins, M. L., and Grunlan, M. A. (2012) The protein resistance of silicones prepared with a PEO-silane amphiphile. *Journal of Materials Chemistry*, **22**, 19540-19546.

90. Rufin, M. A., Barry, M. E., Adair, P. A., Hawkins, M. L., Raymond, J. E., and Grunlan, M. A. (2016) Protein resistance efficacy of PEO-silane amphiphiles: dependence on PEO-segment length and concentration. *Acta Biomaterialia*, **41**, 247-252.

91. Chen, H., Brook, M. A., Chen, Y., and Sheardown, H. (2005) Surface properties of PEO-silicone composites: reducing protein adsorption. *Journal of Biomaterials Science, Polymer Edition*, **16**, 531-548.

92. Chen, H., Brook, M. A., and Sheardown, H. (2004) Silicone elastomers for reduced protein adsorption. *Biomaterials*, **25**, 2273-2282.

93. Chen, H., Zhang, Z., Chen, Y., Brook, M. A. and Sheardown, H. (2005) Protein repellant silicone surfaces by covalent immobilization of

poly(ethylene oxide). *Biomaterials*, **26**, 2391-2399.

94. Rufin, M. A., Gruetzner, J. A., Hurley, M. J., Hawkins, M. L., Raymond, E. S., Raymond, J. E., and Grunlan, M. A. (2015) Enhancing the protein resistance of silicone via surface-restructuring PEO-silane amphiphiles with variable PEO length. *Journal of Materials Chemistry B*, **3**, 2816-2825.

95. Hawkins, M. L., Fay, F., Rehel, K., Linossier, I., and Grunlan, M. A. (2014) Bacteria and diatom resistance of silicones modified with PEO-silane amphiphiles. *Biofouling*, **30**, 247-258.

96. Fay, F., Hawkins, M. L., Rehel, K., Grunlan, M. A., and Linossier, I. (2016) Non-toxic, anti-fouling silicones with variable PEO-silane amphiphile content. *Green Materials*, **4**, 53-62.

97. Galvin, C. J., Dimitriou, M. D., Satija, S. K., Genzer, J. (2014) Swelling of polyelectrolyte and polyzwitterion bushes by humid vapors. *Journal of American Chemical Society*, **136**, 12737–12745.

98. Leng, C., Hung, H., Sun, S., Wang, D., Li, Y., Jiang, S., and Chen, Z. (2015) Probing the surface hydration of nonfouling zwitterionic and PEG materials in contact with proteins. *ACS Applied Materials and Interfaces*, **7**, 16881-16888.

99. Bai, T., Sun, F., Zhang, L., Sinclair, A., Liu, S., Ella-Menye, J., Zheng, Y., and Jiang, S. (2014) Restraint of the differentiation of mesenchymal stem cells by a nonfouling zwitterionic hydrogel. *Angewandte Chemie, International Edition*, **53**, 12729-12734.

100. Yuan, S., Li, Y., Luan, S., Shi, H., and Yan, S. (2016) Infection-resistant styrenic thermoplastic elastomers that can switch from bactericidal capability to anti-adhesion. *Journal of Materials Chemistry B*, **4**, 1081-1089.

101. Jiang, S., and Cao, Z. (2010) Ultralow-fouling, functionalizable, and hydrolyzable zwitterionic materials and their derivatives for biological applications. *Advanced Materials*, **22**, 920-932.

102. Ji, F., Lin, W., Wang, Z., Wang, L., Zhang, J., Ma, G., and Chen, S. (2013) Development of nonstick and drug-loaded wound dressing based on the hydrolytic hydrophobic poly(carboxybetaine) ester analogue. *ACS Applied Materials and Interfaces*, **5**, 10489-10494.

103. Ostuni, E., Grzybowski, B. A., Mrksich, M., Roberts, C. S., and Whitesides, G. M. (2003) Adsorption of proteins to hydrophobic sites on mixed self-assembled monolayers. *Langmuir*, **19**, 1861-1872.

104. Harder, P., Grunze, M., Dahint, R., Whitesides, G. M., and Laibinis, P. E. (1998) Molecular conformation in oligo(ethylene glycol)-terminated self-assembled monolayers on gold and silver surfaces determines their ability to resist protein adsorption. *Journal of Physics Chemistry B*, **102**, 426-436.

105. Noguer, A. C., Olsen, S. M., Hvilsted, S., and Kiil, S. (2016) Long-term stability of PEG-based antifouling surfaces in seawater. *Journal of Coatings and Technology Research*, **13**, 567-575.

106. Cheng, L., Liu, Q., Lei, Y., Lin, Y., and Zhang, A. (2014) The synthesis and characterization of carboxybetaine functionalized polysiloxanes for the preparation of anti-fouling surfaces. *RSC Advances*, **4**, 54372-54381.

107. Zhang, Z., Finlay, J. A., Wang, L., Gao, Y., Callow, J. A., Callow, M. E., and Jiang, S. (2009) Polysulfobetaine-grafted surfaces as environmentally benign ultralow fouling marine coatings. *Langmuir*, **25**, 13516-13521.

108. Aldred, N., Li, G., Gao, Y., Clare, A. S., and Jiang, S. (2010) Modulation of barnacle (*Balanus amphitrite* Darwin) cyprid settlement behavior by sulfobetaine and carboxybetaine methacrylate polymer coatings. *Biofouling*, **26**, 673-683.

109. Goda, T., Konno, T., Takai, M., Moro, T., and Ishihara, K. (2006) Biomimetic phosphorylcholine polymer grafting from polydimethylsiloxane surface using photo-induced polymerization. *Biomaterials*, **27**, 5151-5160.

110. Bodkhe, R. B., Stafslien, S. J., Daniels, J., Cilz, N., Muelhberg, A. J., Thompson, S. E. M., Callow, M. E., Callow, J. A., and Webster, D. C. (2015) Zwitterionic siloxane-polyurethane fouling-release coatings. *Progress in Organic Coatings*, **78**, 369-380.

111. Dundua, A., Franzka, S., and Ulbricht, M. (2016) Improved antifouling properties of polydimethylsiloxane films via formation of polysiloxane/polyzwitterion interpenetrating networks. *Macromolecular Rapid Communications*, **37**, 2030-2036.

112. Shivapooja, P., Yu, Q., Orihuela, B., Mays, R., Rittschof, D., Genzer, J., and Lopez, G. P. (2015) Modification of silicone elastomer surfaces with zwitterionic polymers: short-term fouling resistance and triggered biofouling release. *ACS Applied Materials and Interfaces*, **7**, 25586-25591.

113. Wang, H., Zhang, C., Wang, J., Feng, X., and He, C. (2016) Dual-mode antifouling ability of thiol-ene amphiphilic conetworks: minimally adhesive coatings via the surface zwitterionization. *ACS Sustainable Chemistry & Engineering*, **4**, 3803-3811.

114. Frueh, J., Gai, M. Y., Yang, Z. B., and He Q. (2014) Influence of polyelectrolyte multilayer coating on the degree and type of biofouling in freshwater environment. *Journal of Nanoscience and Nanotechnology*, **14**, 4341-4350.

115. Kuliasha, C. A., Finlay, J. A., Franco, S. C., Clare, A. S., Staslien, S. J., and Brennan, A. B. (2017) Marine anti-biofouling efficacy of amphiphilic poly(coacrylate) grafted PDMSe: effect of graft molecular weight. *Biofouling*, **33**, 252-267.

116. Bodkhe, R. B., Stafslien, S. J., Cilz, N., Daniels, J., Thompson, S. E. M., Callow, M. E., Callow, J. A., and Webster, D. C. (2012) Polyurethanes with amphiphilic surfaces made using telechelic functional PDMS having orthogonal acid functional groups. *Progress in Organic*

Coatings, **75**, 38-48.

117. Webster, D. C., and Bodkhe, R. B. (2015) Functionalized Silicones With Polyalkylene Oxide Side Chains, patent US9169359B2.

118. Galhenage, T. P., Webster, D. C., Moreira, A. M. S., Burgett, R. J., Stafslien, S. J., Vanderwal, L., Finlay, J. A., Franco, S. C., and Clare, A. S. (2017) Poly(ethylene) glycol-modified, amphiphilic, siloxane-polyurethane coatings and their performance as fouling-release surfaces. *Journal of Coatings and Technology Research,* **14**, 307-322.

119. Jiang, J., Fu, Y., Zhang, Q., Zhan, X., and Chen, F. (2017) Novel amphiphilic poly(dimethylsiloxane) based polyurethane networks tethered with carboxybetaine and their combined antibacterial and anti-adhesive property. *Applied Surface Science*, **412**, 1-9.

120. Cavalli, S., Albericio, F., and Kros, A. (2010) Amphiphilic peptides and their cross-disciplinary role as building blocks for nanoscience. *Chemical Society Reviews*, **39**, 241-263.

121. Chen, S., Cao, Z., and Jiang, S. (2009) Ultra-low fouling peptide surfaces derived from natural amino acids. *Biomaterials*, **30**, 5892-5896.

122. Ye, H., Wang, L., Huang, R., Su, R., Liu, B., Qi, W., and He, Z. (2015) Superior antifouling performance of a zwitterionic peptide compared to an amphiphilic, non-ionic peptide. *ACS Applied Materials and Interfaces*, **7**, 22448-22457.

123. Schneider, M., Tang, Z., Richter, M., Marschelke, C., Forster, P., Wegener, E., Amin, I., Zimmermann, H., Scharnweber, D., Braun, H.-G., Luxenhofer, R., and Jordan, R. (2016) Patterned polypeptoid brushes. *Macromolecular Bioscience*, **16**, 75-81.

124. Statz, A. R., Barron, A. E., and Messersmith, P. B. (2008) Protein, cell and bacterial fouling resistance of polypeptoid-modified surfaces: effect of side-chain chemistry. *Soft Matter*, **4**, 131-139.

125. Statz, A. R., Meagher, R. J., Barron, A. E., and Messersmith, P. B. (2005) New peptidomimetic polymers for antifouling surfaces. *Journal of American Chemical Society*, **127**, 7972-7973.

126. Patterson, A. L., Wenning, B., Rizis, G., Calabrese, D. R., Finlay, J. A., Franco, S. C., Zuckermann, R. N., Clare, A. S., Kramer, E. J., Ober, C. K., and Segalman, R. A. (2017) Role of backbone chemistry and monomer sequence in amphiphilic oligopeptide- and oligopeptoid-functionalized PDMS- and PEO-based block copolymers for marine antifouling and fouling release coatings. *Macromolecules*, **50**, 2656-2667.

127. Calabrese, D. R., Wenning, B., Finlay, J. A., Callow, M. E., Callow, J. A., Fischer, D., and Ober, C. K. (2015) Amphiphilic oligopeptides grafted to PDMS-based diblock copolymers for use in antifouling and fouling release coatings. *Polymers for Advanced Technologies*, **26**, 829-836.

128. Van Zoelen, W., Buss, H. G., Ellebracht, N. C., Lynd, N. A., Fischer, D.

A., Finlay, J., Hill, S., Callow, M. E., Callow, J. A., Kramer, E. J., Zuckermann, R. N., and Segalman, R. A. (2014) Sequence of hydrophobic and hydrophilic residues in amphiphilic polymer coatings affects surface structure and marine antifouling/fouling release properties. *ACS Macro Letters*, **3**, 364-368.

129. Van Zoelen, W., Zuckermann, R. N., and Segalman, R. A. (2012) Tunable surface properties from sequence-specific polypeptoid-polystyrene block copolymer thin films. *Macromolecules*, **45**, 7072-7082.

130. Leng, C., Buss, H. G., Segalman, R. A., and Chen, Z. (2015) Surface structure and hydration of sequences specific amphiphilic polypeptoids for antifouling/fouling release applications. *Langmuir*, **31**, 9306-9311.

131. Calabrese, D. R., Wenning, B. M., Buss, H., Finlay, J. A., Fischer, D., Clare, A. S., Segalman, R. A., and Ober, C. K. (2017) Oligopeptide-modified hydrophobic and hydrophilic polymers as antifouling coatings. *Green Materials*, **5**, 31-43.

132. Lei, Y., Lin, Y., and Zhang, A. (2017) The synthesis and protein resistance of amphiphilic PDMS-*b*-(PDMS-*g*-cysteine) copolymers. *Applied Surface Science*, **419**, 393-398.

133. Olsen, S. M., Pedersen, L. T., Laursen, M. H., Kiil, S., and Dam-Johansen, K. (2007) Enzyme-based antifouling coatings: a review. *Biofouling*, **23**, 369-383.

134. Cordeiro, A. L., Hippius, C., and Werner C. (2011) Immobilized enzymes affect biofilm formation. *Biotechnology Letters*, **33**, 1897-1904.

135. Tasso, M., Conlan , S. L., Clare , A. S., and Werner, C. (2012) Active enzyme nanocoatings affect settlement of *Balanus amphitrite* barnacle cyprids. *Advanced Functional Materials*, **22**, 39-47.

136. Christie, A. O., Evans, L. V., and Shaw, M. (1970) Studies on the ship-fouling alga enteromorpha II. The effect of certain enzymes on the adhesion of zoospores. *Annals of Botany*, **34**, 467-482.

137. Kim, Y., Dordick, J., and Clark, D. (2001) Siloxane-based biocatalytic films and paints for use as reactive coatings. *Biotechnology and Bioengineering*, **72**, 475-482.

138. John, H. Z., Jackson, K., Bian, C., Burt, H. M., and Chiao, M. (2015) Enzyme-modified hydrogel coatings with self-cleaning abilities for low fouling PDMS devices. *Advanced Materials Interfaces*, **2**, 1500154.

139. Olsen, S. M., and Yebra, D. M. (2013) Polysiloxane-Based Fouling Release Coats Including Enzymes, WO2013000477A1.

140. Pavlovic, D., Lafond, S., Margaillan, A., and Bressy, C. (2016) Facile synthesis of graft copolymers of controlled architecture. Copolymerization of fluorinated and non-fluorinated poly(dimethylsiloxane) macromonomers with trialkylsilyl methac-

rylates using RAFT polymerization. *Polymer Chemistry*, **7**, 2652-2664.

141. Duong, T. H., Briand, J. F., Margaillan, A., and Bressy, C. (2015) Polysiloxane-based block copolymers with marine bacterial antiadhesion properties. *ACS Applied Materials and Interfaces*, **7**, 15578-15586.

142. Lejars, M., Margaillan, A., and Bressy, C. (2013) Well-defined graft copolymers of tert-butyldimethylsilyl methacrylate and poly(dimethylsiloxane) macromonomers synthesized by RAFT polymerization. *Polymer Chemistry*, **4**, 3282-3292.

143. Vaterrodt, A., Thallinger, B., Daumann, K., Koch, D., Guebitz, G. M., and Ulbricht, M. (2016) Antifouling and antibacterial multifunctional polyzwitterion/enzyme coating on silicone catheter material prepared by electrostatic layer-by-layer assembly. *Langmuir*, **32**, 1347-1359.

144. Voo, Z. X., Khan, M., Narayanan, K., Seah, D., Hedrick, J. L., and Yang, Y. Y. (2015) Antimicrobial/antifouling polycarbonate coatings: Role of block copolymer architecture. *Macromolecules*, **48**, 1055-1064.

145. Majumdar, P., Lee, E., Patel, N., Stafslien, S. J., Daniels, J., and Chisholm, B. J. (2008) Development of environmentally friendly, antifouling coatings based on tethered quaternary ammonium salts in a crosslinked polydimethylsiloxane matrix. *Journal of Coatings and Technology Research*, **5**, 405-417.

146. Majumdar, P., Crowley, E., Htet, M., Stafslien, S. J., Daniels, J., VanderWal, L., and Chisholm, B. J. (2011) Combinatorial materials research applied to the development of new surface coatings XV: an investigation of polysiloxane anti-fouling/fouling-release coatings containing tethered quaternary ammonium salt groups. *ACS Combinatorial Science*, **13**, 298-309.

147. Majumdar, P., Mayo, B., Kim, J., Gallagher-Lein, C., Lee, E., Gubbins, N., and Chisholm, B. J. (2010) The utilization of specific interactions to enhance the mechanical properties of polysiloxane coatings. *Journal of Coatings and Technology Research*, **7**, 239-252.

148. Nurdin, N., Helary, G., Sauvet, G. (1993) Biocidal polymers active by contact. II. Biological evaluation of polyurethane coatings with pendant quaternary ammonium salts. *Journal of Applied Polymer Science*, **50**, 663-670.

149. Hazziza-Laskar, J., Helary, G., Sauvet, G. (1995). Biocidal polymers active by contact. IV. Polyurethanes based on polysiloxanes with pendant primary alcohols and quaternary ammonium groups. *Journal of Applied Polymer Science*, **58**, 77-84.

150. Majumdar, P., Lee. E., Patel, N., Ward, K., Stafslien. S. J., Daniels, J., Chisholm, B. J., Boudjouk, P., Callow, M. E., Callow, J. A., and Thompson, S. E. (2008) Combinatorial materials research applied

to the development of new surface coatings IX: an investigation of novel antifouling/fouling-release coatings containing quaternary ammonium salt groups. *Biofouling*, **24**, 185-200.

151. Xie, Q., Ma, C., Liu, C., Ma, J., and Zhang, G. (2015) Poly(dimethylsiloxane) based polyurethane with chemically attached antifoulants for durable marine anti-biofouling. *ACS Applied Materials and Interfaces*, **7**, 21030-21037.

152. Ma, J., Ma, C., Yang, Y., Xu, W., and Zhang, G. (2014) Biodegradable polyurethane carrying antifoulants for inhibition of marine biofouling. *Industrial and Engineering Chemistry Research*, **53**, 12753-12759.

153. Noguer, A. C., Kiil, S., Dam-Johansen, K., Hvilsted, S., and Olsen, S. M. (2016). *Experimental Investigation of the Behaviour and Fate of Block Copolymers in Fouling-release Coating*, PhD Thesis, Technical University of Denmark, Denmark.

154. Lines, R., Willimans, D. N., and Turri, S. (2003) Antifouling Coating Composition Comprising a Fluorinated Resin, patent US2003161962.

155. Reynolds, K. J., and Tyson, B. V. (2016) Anti-fouling Compositions with a Fluorinated Oxyalkylene-Containing Polymer or Oligomer, patent EP2961805B1.

156. Tyson, B. V., and Reynolds, K. J. (2016) Fouling-Resistant Composition Comprising Sterols and/or Derivatives Thereof, patent US9388316B2.

157. Thorlaksen, P. C. W., Blom, A., and Bork, U. (2015) Novel Fouling Control Coating Compositions, patent EP2516559B1.

158. Thorlaksen, P. C. W. (2017) Fouling Control Coating Compositions Comprising Polysiloxane and Pendant Hydrophilic Oligomer/Polymer Moieties, patent EP2726558B1.

159. Watermann, B., Berger, H., Sonnichsen, H., and Willemsen, P. (1997) Performance and effectiveness of nonstick coatings in seawater. *Biofouling*, **11**, 101-118.

160. Amidaiji, K., Tashiro, S., and Sakamoto, T. (2009) Curable Composition, Antifouling Coating Composition, Antifouling Coating Film, Base With Antifouling Coating Film, and Method for Preventing Fouling on Base, patent EP2103655A1.

7

Anti-corrosion Behavior of Layer-by-Layer Coatings of Crosslinked Chitosan and Poly(vinyl butyral) on Carbon Steel

Gisha Elizabeth Luckachan and Vikas Mittal*,**

Department of Chemical Engineering, The Petroleum Institute (part of Khalifa University of Science and Technology), Abu Dhabi, UAE

*Corresponding author: vik.mittal@gmail.com
**Current address: Bletchington, Wellington County, Australia

7.1 Introduction

Corrosion of metals is one of the main destruction processes resulting in huge economic losses, especially in the petroleum, aerospace and automotive industries. Application of organic coatings is most widely used method for the passive corrosion protection of metallic structures [1-3]. Recently, much research effort has been focused on the coatings using renewable resources such as chitin, chitosan, cellulose, polylactide, etc. due to advantages such as cost effectiveness, low toxicity, inherent biodegradability and environment friendliness [4,5]. Among these, chitosan (Ch), a linear polyamine consisting primarily of β linked 2-amino-2-deoxy-β-d glucopyranose units, has been widely used for the development of anti-corrosion coatings because of its unique combination of properties such as antimicrobial activity, chemical stability, biocompatibility, and good film forming properties [6,7]. Chitosan can adhere to negatively charged surfaces, therefore, spontaneously adsorbing on metal or oxide surfaces. It can form complexes with metal ions, and gels with polyanions. The hydroxyl and amine groups on chitosan are reactive and can be used to generate different functional derivatives with desired properties needed for effective corrosion protection [8-13]. The controllable release of the active compounds introduced to the chitosan films is also possible, thus, making these films attractive for applications when active corrosion protection is required [6,12,14,15].

Marine Coatings and Membranes, edited by Vikas Mittal
© 2019 Central West Publishing, Australia

A critical drawback in using chitosan as a corrosion-preventing barrier is its absorption of large amounts of moisture from the atmosphere, thus, forming hydrogels. This transformation not only leads to biodegradation of the film, but also allows moisture to infiltrate easily into the film, causing its failure as a protective coating [12,16]. One strategy to overcome this drawback is to associate chitosan with a moisture-resistant polymer. Sugama and Cook [12] generated water-insoluble chitosan biopolymer coatings on aluminum by grafting synthetic poly(itaconic acid) polymer onto the linear chitosan chains. Dextrin modification of chitosan has also been performed by the same group in an aim to decrease the hydrophilicity and improve the bond strength of chitosan coatings [17]. Most of the organic compounds involved in corrosion inhibition contain oxygen, nitrogen and/or sulfur groups, which adsorbed on the metallic surface thereby blocking the active corrosion sites. Since chitosan is naturally rich in such hydroxyl and amino groups, it has a strong potential as corrosion inhibitor, however, a little has been reported about the corrosion inhibition using chitosan. El-Haddad [18] and Umoren et al. [19] reported inhibition of copper corrosion and mild steel corrosion respectively in acid medium using chitosan. Aminothiourea modified chitosan was used by Manlin et al. [20] for the protection of steel in acetic acid. Mohammed and Fekry [21] reported the protection of steel by chitosan crotonaldehyde Schiff's base in salt solution.

In the current study, stable chitosan coatings on mild carbon steel were generated by layer-by-layer (lbl) addition of chitosan and hydrophobic polymer like poly(vinyl butyral) (PVB) along with analyzing the anti-corrosion performance in 0.3 M salt solution. Glutaraldehyde was used as the crosslinking agent for chitosan in order to reduce the water absorption of chitosan and to enhance coating stability. PVB was selected due its adhesion property along with hydrophobic nature. Lbl coating method was employed to generate a uniform coverage of substrate with chitosan instead of composite film coatings of PVB and chitosan (a method commonly used to improve the performance of hydrophilic polymers). Thus way, PVB over-coating was enabled to prevent chitosan from direct contact with corrosive environment which improved the coating stability and corrosion performance as well. Corrosion inhibition of chitosan in the lbl coatings was analyzed using electrochemical impedance spectroscopy (EIS) and Tafel measurements. Role of small amount of inorganic fillers like graphene and vermiculite in the chitosan layer on the corrosion performance was also studied.

7.2 Experimental

7.2.1 Materials

PVB, hydrochloric acid and acetic acid were purchased from Aldrich, Germany and were used as received. PVB (trade name of Butvar B98) had molecular weight of 40000-70000 g/mol and specific gravity of 1.1 at 23 °C. Butyral content (expressed as % polyvinyl butyral) was 80% in the PVB resin used. Chitosan flakes were procured from Bio21, Thailand. Chitosan had a molecular weight of 180000 g/mol and 90% degree of deacetylation (DD). Glutaraldehyde was purchased from Fisher Chemicals. Russian vermiculite with a chemical composition of $(Mg,Al,Fe)_3(Al,Si)_4O_{10}(OH)_2Mg_x(H_2O)_n$ was obtained from Thermax, Greinsfurth, Austria. It was received from the supplier as flakes and was subsequently wet ground to the desired particle diameter (partially spherical stacks with average particle size of 5 μm). The grinding operation was performed keeping in mind that the lateral dimensions of the platelets are not milled, however, the thickness of the stacks is reduced so as to generate platelets with high aspect ratio. In the as-supplied form, it contained Mg^{2+} ions on the surface. Single layer graphene platelets were procured from Angstron Materials, USA.

7.2.2 Lbl Coatings of Chitosan and PVB

0.5 mg chitosan was dissolved in 10 ml of 2% acetic acid. This mixture was stirred for 24 h to ensure complete dissolution of chitosan in acetic acid. 0.5 mg of PVB was dissolved in 10 ml of methanol. Lbl coating was applied on 5 cm x 3 cm x 2 cm carbon steel coupons by using a dip coater. A layer of PVB was first applied, followed by two layers of chitosan. After drying at 75 °C for 2 h, a final PVB layer was applied. Final coating was cured for 2 h at 90 °C.

7.2.3 PVB_Ch/x%Glu_PVB Coatings

To generate coatings with crosslinked chitosan, 5 ml of x% (where x = 1%, 5% and 10%) glutaraldehyde solution in water was added to 0.5 mg of chitosan dissolved in 10 ml of 2% acetic acid and stirred for 15 min. Carbon steel coupons coated with PVB as first layer were dip coated two times in the chitosan-glutaraldehyde solution. After drying at 75 °C for 2 h, a final layer of PVB were coated. Final coatings were cured at 90 °C for 2 h. This way, three different coatings were prepared by changing the glutaraldehyde content to 1%, 5% and 10% in PVB_Ch/x%Glu_PVB coating. In order to prepare PVB_Ch/1%Glu/5%Gr_PVB composite coating, 0.025 g of graphene was added to 0.475 g of chitosan in 10 ml of 2% acetic acid solution and sonicated for 24 h. After sonication, 5 ml of 1% glutaraldehyde solution was added to this mixture and dip coated on the carbon steel coupons using the lbl approach described above. PVB_Ch/1%Glu/5%Ver_PVB coating were prepared by following the same procedure using 0.025 g vermiculite instead of graphene nano-platelets.

7.2.4 Immersion Test

The edges of the coated coupons were sealed using Nippon epoxy primer. After 24 h drying at RT, samples were immersed in 0.3 M aerated salt solution for standard corrosion analysis.

7.2.5 Electrochemical Measurements

EIS measurements and Tafel plots were carried out at RT in a three-electrode corrosion cell consisting of a saturated calomel reference electrode, a platinum counter electrode and lbl coated steel coupons as the working electrode. 0.3 M salt solution was used as electrolyte. All measurements were performed on computerized electrochemical analyzer (supplied by BioLogic, France). 1 cm^2 area of working electrode was exposed to electrolyte and impedance measurements were performed as a function of open circuit potential (EOCV). The selected frequency range was typically from 10^5 to 10^{-2} Hz at AC amplitude of 10 mV. The impedance plots were fitted using an equivalent circuit (given in Figure 7.5), where pure capacitances were replaced by constant-phase elements (CPE). The software employed for the fitting process was EC-Lab V10.39. The capacitance values of the different elements in the equivalent circuit were calculated using the following equation:

$$C = Y_0 \left(\omega_{max}\right)^{n-1}$$

where ω_{max} is the frequency at which the imaginary impedance reaches a maximum for the respective time constant [22,23]. Tafel plots were recorded at a scan rate of 0.166 mV/s.

7.2.6 Fourier Transformed Infrared Spectroscopy and Raman Spectroscopy

Structural characterization of lbl coatings of chitosan and PVB was carried out using a Bruker VERTEX 70 FTIR spectrometer attached with a DRIFT accessory. IR acquisition was achieved between 4000 cm^{-1} - 370 cm^{-1} using OPUS software at 4 cm^{-1} resolution. 128 scans were used for each acquisition. Raman spectra were recorded using LabRAM HR spectrometer (Horiba Jobin Yvon). Laser light from He/Ne source with wavelength of 633 nm was used for excitation. A long working distance objective with magnification 50x was used to collect the scattered light as well as to focus the laser beam on the sample surface.

7.3 Results and Discussion

In the current study, coatings of chitosan were generated by layer-by-layer addition of chitosan and PVB on mild carbon steel surface. Final coatings comprised of one layer of PVB followed by two layers of chitosan and final layer of PVB; denoted as PVB_Ch_PVB. Chitosan layer in the lbl coating were also modified further by crosslinking with glutaraldehyde which is denoted as PVB_Ch/x%Glu_PVB. In addition, the chitosan coatings were also reinforced by addition of 5 wt% of functional fillers like graphene and vermiculite, which are denoted as PVB_Ch/1%Glu/5%Gr_PVB and PVB_Ch/1%Glu/5%Ver_PVB.

7.3.1 Structure of lbl Coatings of Chitosan and PVB

Structure of lbl coatings was analyzed using DRIFT spectroscopy. Figure 7.1 shows the IR spectra of chitosan and PVB, along with PVB_Ch and PVB_Ch/1%Glu coatings without a PVB top layer. Properties of PVB are characterized by the butyral units as well as hydroxyl and acetal groups [24,25]. C-O-C-O-C stretching vibrations of butyral units showed bands at 1167 cm^{-1} and 1012 cm^{-1} in the PVB spectrum. Less intense peak at 1736 cm^{-1} corresponded to C=O stretching vibration and peaks at 1006 cm^{-1} and 1242 cm^{-1} resulted from C-O-C stretching vibrations of acetate groups. Intense broad band at 3470 cm^{-1} was assigned to –OH stretching vibrations of hydroxyl group, the quantity of which determines the crosslinking capacity of PVB. C-H bending and stretching vibrations of saturated –CH, –CH$_2$ and –CH$_3$ were observed at regions of 2720 cm^{-1} to 2850 cm^{-1} and 1500 cm^{-1} to

1300 cm-1 respectively [24,25]. Raw chitosan spectrum exhibited an intense broad band at 3477 cm-1 corresponding to axial stretching of O-H and N-H bonds. Bands located at 1667 cm-1 and 1597 cm-1 were assigned to the amide 1 and amide 11 vibrations respectively, small hump at 1543 cm-1 corresponded to protonated amine, whereas bands at 1425 cm-1 and 1384 cm-1 resulted from the bending vibrations of methyl and methylene groups respectively. The absorption bands in the range of 1000 cm-1 to 1200 cm-1 corresponded to the polysaccharide backbone including the glycosidic bonds, C-O and C-O-C stretching vibrations [14,26-28].

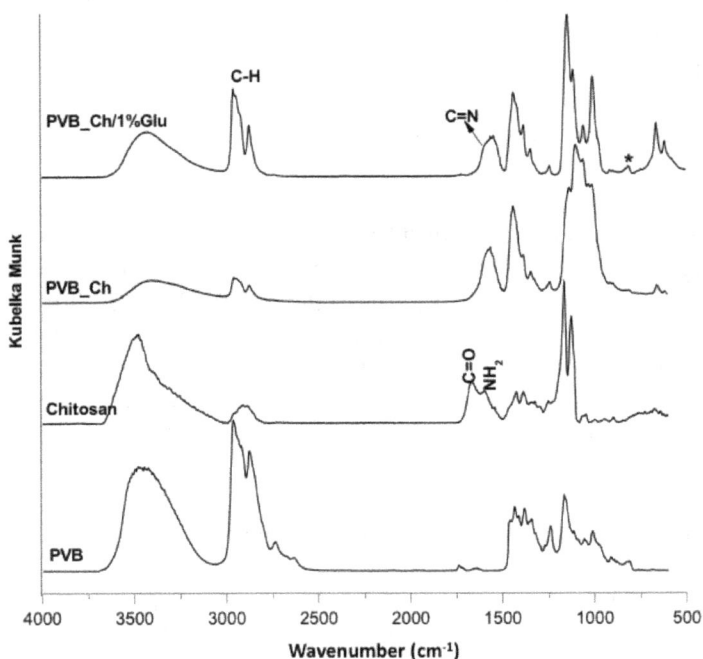

Figure 7.1 IR spectra of raw PVB, raw chitosan, PVB_Ch coating and PVB_Ch/1%Glu coating.

PVB and chitosan polymers exhibited significant changes in the IR spectra of PVB_Ch coating. Hydroxyl and amine vibrations of PVB and chitosan at around 3470cm-1 became less intense and amide carbonyl vibration of chitosan at 1640 cm-1 disappeared almost completely. Taking these changes into account, it can be assumed that PVB hydroxyl and chitosan amide carbonyl groups reacted to result in cross-linking between the two polymers. Not only the shifting of hydroxyl vibrations to 3400 cm^{-1}, but also the broadening of this band indicated further the chances of hydrogen bond interaction between the two polymers. Incorporation of glutaraldehyde in the chitosan layer introduced a new shoulder band at 1600 cm^{-1}, which was attributed to valence vibrations of C=N bond of azomethin group formed by the crosslinking reaction of chitosan with glutaraldehyde [26,29]. Moreover, the increase in the number of methylene groups in the crosslinked product, attributed to the glutaraldehyde methylene groups, resulted in the increased intensity of C-H vibrations in the region of 3000 cm^{-1} - 2850 cm^{-1}. Hydrogen bonding between chitosan chains became weak by the crosslinking with glutaraldehyde that was observed from the shift of hydroxyl vibrations to a higher wavenumber of 3433 cm^{-1}, evolution of sharp and well resolved peaks attributed to amine and polysaccharide backbone vibrations below 1600 cm^{-1} and formation of two separate distinguishable bands at 1560 cm^{-1} and 1543 cm^{-1} corresponding to –NH- bridging vibrations and protonated amine respectively [30]. These spectroscopic results indicated a crosslinking reaction between PVB, chitosan and glutaraldehyde in the lbl coatings. Based on these results, proposed crosslinking mechanism is shown in Figure 7.2.

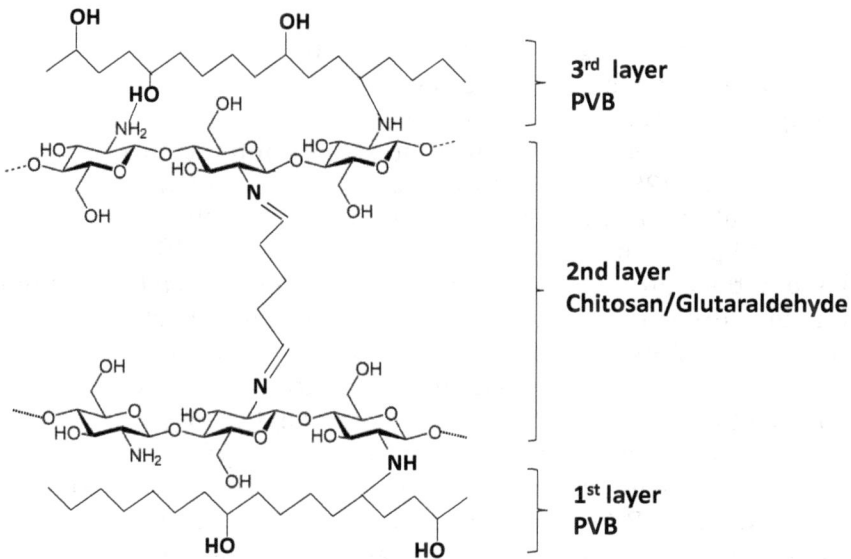

Figure 7.2 Schematic representation of crosslinking of chitosan with PVB and glutaraldehyde in the PVB_Ch/x%Glu_PVB coatings.

7.3.2 Anti-corrosion Performance of Chitosan

In order to analyze the protective behavior of pure chitosan, PVB and chitosan coatings of 3 μm thickness were prepared separately, and EIS experiments were conducted on these coatings along with bare steel in 0.3 M salt solution. The Bode and phase plots of all three samples measured after 2 h immersion in 0.3 M salt solution, given in Figure 7.3a, showed a single time constant at intermediate frequencies (10^0 to 10^2 Hz). This relaxation process is associated with the double layer capacitance of the electrolyte at the metal surface indicating the corrosion process. However, low frequency Z-modulus, related to the corrosion protection of the coatings, was higher for chitosan coating than PVB coating which showed the anti-corrosion property of chitosan [31]. It was also observed in the Nyquist plot as increased diameter of the semicircle (Figure 7.3b). It has been reported that as chitosan is naturally functionalized with amine and hydroxyl groups, these groups can bind or ionically interact with steel via the lone pairs of electrons on the O and N atoms which provide inhibition to steel corrosion [19,32]. Such an interaction of chitosan functional groups with steel surface has also been observed in the IR spectra of lbl coatings given in Figure 7.1. Polysaccharide backbone vibrations of chitosan at 1163 cm^{-1} and 1125 cm^{-1} changed to a broad band with multiple peaks at the region of 1200 cm^{-1} - 1000 cm^{-1} in the spectra of PVB_Ch coating which showed the interaction of chitosan hydroxyl groups with steel surface. Free electron doublet on the nitrogen atoms of amine groups are the main sites for the interaction of chitosan with iron ions [28]. However, this interaction was not distinguishable because of the overlapping of bands assigned to -NH- bridge vibrations and protonated amine at 1559 cm^{-1}. However, NH deformation band in the chitosan spectra observed at 680 cm^{-1} as a broad hump due to hydrogen bonding formed a well-defined peak at 667 cm^{-1} in the spectra of PVB_Ch coating, which is attributed to the interaction of chitosan amine groups with metal surface. Chitosan crosslinking with glutaraldehyde reduced intermolecular hydrogen bonding of chitosan chains, and, thus, released more free hydroxyl and amine groups suitable for chelation with metal ions. It was observed in the IR spectra of PVB_Ch/1%Glu coating, where NH deformation of chitosan formed a sharp intense band at 663 cm^{-1} and well separated intense hydroxyl bands in 1200 cm^{-1} to 1000 cm^{-1} region. A new band at 812 cm^{-1} attributed to νNH_2 and ρNH_2 also evolved from the interaction of amine group with metal ions [28]. All these changes lead to the probability of chitosan interaction with metal ions through hydroxyl and amine groups which might contribute to the enhanced corrosion protection of underlying metal surface in the lbl coatings. However, corrosion protection of chitosan coatings did not last for long time because of the relatively high affinity of chitosan towards water which opens the pathways for the diffusion of electrolyte ions down to the metal surface.

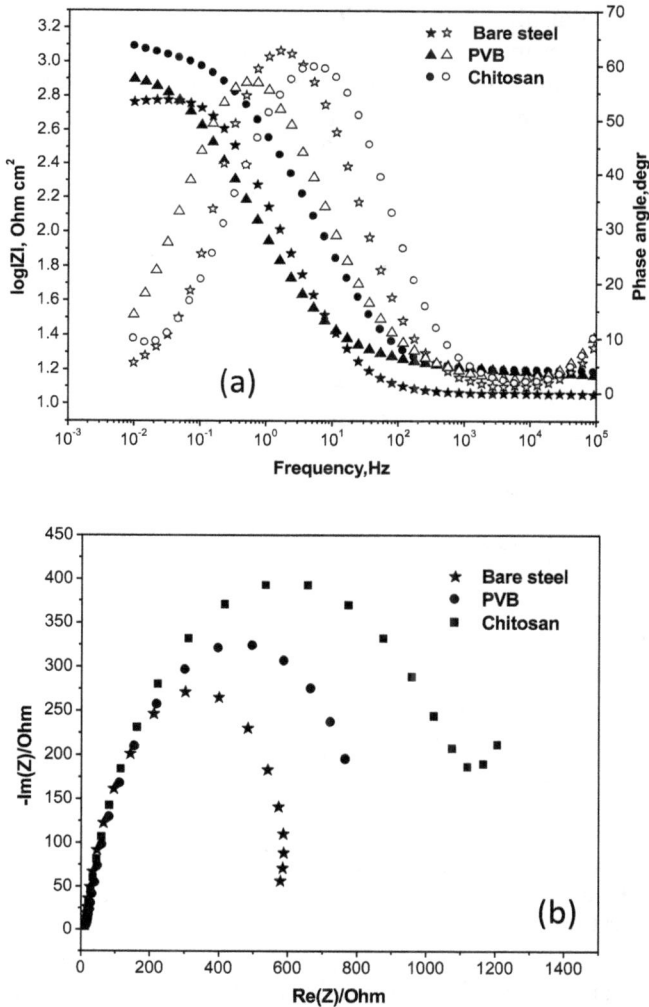

Figure 7.3 Bode & Phase plots (a) and Nyquist plots (b) of bare steel, PVB coating and chitosan coating after 2 h immersion in 0.3 M salt solution.

7.3.3 Corrosion Protection Properties of lbl Coatings of PVB and Chitosan

Sandwiching of chitosan in between hydrophobic PVB coatings was achieved to improve the adhesive strength as well as to avoid the direct contact of chitosan with the corrosive medium. The resulting PVB_Ch_PVB coatings were modified further with the incorporation of different percentages of glutaraldehyde in the chitosan layer and coatings of overall 12 μm thickness were used for anti-corrosion analysis. Corrosion protection of PVB_Ch_PVB coatings in 0.3 M salt solution was monitored by EIS measurements at different times of immersion, which are represented graphically in the Bode and Nyquist plots given in Figure 7.4. The impedance at lowest frequency

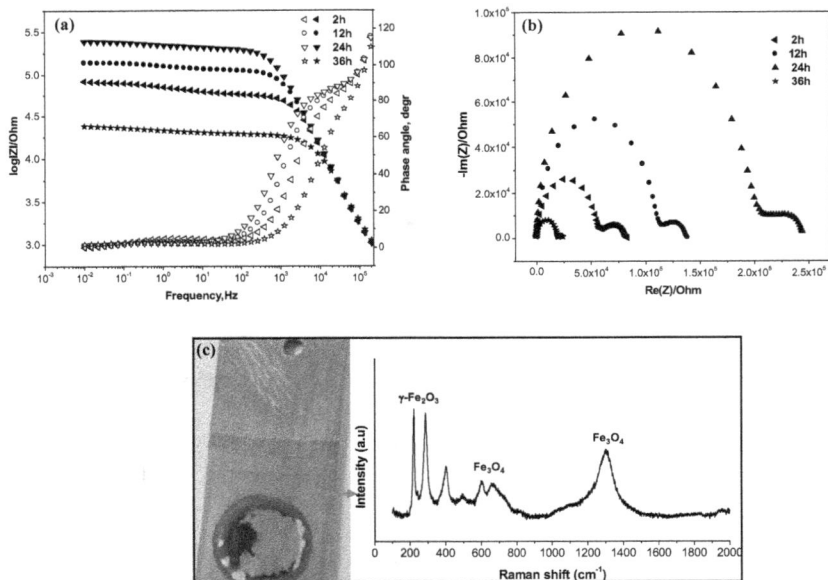

Figure 7.4 (a) Bode and phase plots and (b) Nyquist plots of PVB_Ch_PVB coating at different times of immersion in 0.3 M salt solution; (c) Optical photograph and Raman spectra of PVB_Ch_PVB coated steel surface after 24 h EIS and film removal.

(10^{-2} Hz) in the Bode plot obtained after 2 h immersion was higher by approximately two orders of magnitude than pure chitosan coating. The Bode plot contained two well defined time constants, which are clearly described in the Nyquist plot with two semicircles. Generally,

EIS plots of coatings having barrier property show a time constant at high frequency region [6]. Therefore, the time constant at high frequency region (10^5 Hz) in the Bode plot represented the coating response, which might have resulted from the additional barrier provided by the PVB over-coating. The second time constant at low frequency (10^1 Hz to 10^{-1} Hz) was attributed to the charge/transfer process at the metal electrolyte interface. As the time of immersion increased, low frequency impedance in the Bode plot continued to increase and reached a maximum at 24 h and subsequently started to decrease. Such an electrochemical behavior could be described in terms of an equivalent circuit depicted in Figure 7.5. The symbol R_{sol} represents the solution resistance of the bulk electrolyte between the reference and working electrode, R_p is the pore resistance of the lbl coating, R_{ct} is the charge transfer resistance, and Q_{dl} and Q_c are the constant phase elements (CPEs). The double layer does not behave as

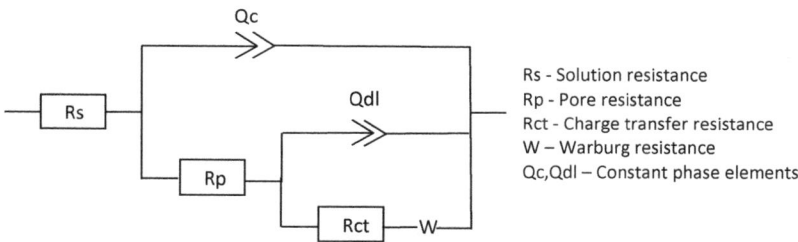

Figure 7.5 Equivalent circuit used for EIS modeling.

an ideal pure capacitor in the presence of the dispersing effect, so a constant phase element (Q_{dl}) is used as substitute for the capacitor. Dispersing effect of double layer was introduced in the equivalent circuit by the addition of a Warburg element (W). The constant phase element Q_c may be caused by the special property of the lbl coating. There is a chance for interface between the PVB and chitosan layers which contributed to an existence of another interface capacitance-like element reflected by electrical signals. As this interface does not behave as an ideal capacitor, so a constant phase element (Q_c) was used instead. The value of CPE is a function of the angular frequency, ω, and its phase is independent of the frequency. Its admittance and impedance are respectively expressed as

$$Y_{CPE} = Y_0(j\omega)^n$$
$$Z_{CPE} = 1/Y_0(j\omega)^{-n}$$

where Y_0 is the magnitude of the CPE, ω is the angular frequency, and n is the exponential term of the CPE which can vary between 1 for pure capacitance and 0 for a pure resistor [33]. The electronic parameters obtained from fitting the experimental EIS using the above equivalent circuit are listed in Table 7.1.

Table 7.1 EIS parameters of PVB_Ch_PVB coating at different time of immersion in 0.3 M salt solution

Coating	Exposure time	R_p	C_c	Q_c	
	h	Ωcm^2	Fcm^2	$Y_{0\,(F.s^{\wedge}.n-1)}$	n
PVB_Ch_PVB	2	0.05e6	1.68e-4	2.4e-9	0.92
	12	0.11e6	1.65e-4	2.0e-9	0.93
	24	0.20e6	1.69e-4	1.9e-9	0.94
	36	0.19e5	3.284e-3	4.0e-6	0.88

Coating	Exposure time	R_{ct}	C_{dl}	Q_{dl}	
	h	Ωcm^2	Fcm^2	$Y_{0\,(F.s^{\wedge}.n-1)}$	n
PVB_Ch_PVB	2	25930	3.00e-2	11.5e-6	0.57
	12	27240	3.05e-2	12.7e-6	0.60
	24	39637	3.21e-2	7.38e-6	0.62
	36	418	5.85e-2	2.05e-2	0.07

During the immersion in 0.3 M salt solution, pore resistance (R_p) of PVB_Ch_PVB coatings exhibited an increasing trend, and, after 24 h, it reached a value four times higher than the initial value. Since pore resistance is related to the resistance of the electrolyte in pores, cracks and pits in the coating and reflects barrier property of the coating, it can be suggested that water permeability of PVB_Ch_PVB coatings decreased significantly during the initial exposure time [34,35]. At the same time, coating capacitance (C_c) was observed to remain constant over 24 h which indicated that water entered the coating was dispersed throughout the coating instead of localization at the metal/coating interface as free electrolyte [35]. Further immersion in 0.3 M salt solution increased C_c and decreased R_p values indicating the diffusion of more water/ions into the coating that led to complete degradation of the coating [35]. The appearance of second loop in the Nyquist plot of PVB_Ch_PVB coatings and the corresponding fitting parameters (R_{ct} and C_{dl}) showed that electrochemical processes started on the metal/solution interface at the early stages of immersion. Corrosion on steel surface involves several oxidation and reduction process given in the equations below [36]:

$$Fe \rightarrow Fe^{2+} + 2e^-$$
$$Fe^{2+} \rightarrow Fe^{3+} + 1e^-$$
$$O_2(g) + 2H_2O + 4e^- \rightarrow 4OH^-$$
$$2Fe^{2+}(aq) + O_2(g) + H_2O \rightarrow 2FeOOH + H_2O$$

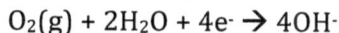

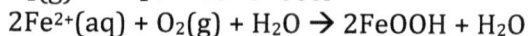

As per the equations, a passive oxide layer is formed firstly on steel surface, which is supposed to prevent the underlying metal from further corrosion, similar to other metals like aluminum, copper, etc. [37]. However, in the case of steel, these oxides are highly porous, which provide a path for the migration of electrolytes down to the metal surface and hence further corrosion occurs below the passive oxide layer. Hence, by stabilizing this oxide layer, it would be possible to slow down the corrosion process on metal surface. Such stabilization of passive oxide layer has been observed in the chitosan coatings. Chitosan is well known for its ability to form stable complexes with transition metal ions. Amino and amide nitrogen, alcoholic and etheric oxygen donor atoms in chitosan can act as a potential binding sites for Fe^{2+} and Fe^{3+} ions [27]. It has been reported that iron ion chelates with 2 moles of amino groups and four moles of oxygen atoms in chitosan, and when pH of the medium changes, it oxidizes the iron cations to form chitosan stabilized iron oxides [28]. Therefore, the changes observed in the PVB_Ch_PVB coatings during corrosion analysis indicate that as water/ions reach the metal surface, these facilitate the corrosion process involved in step 1 and 2. The formed Fe^{2+} and Fe^{3+} ions on the metal chelated with chitosan through free amino and hydroxyl groups, which acted as the nucleation sites for the further oxidation. Finally, a passive layer of iron oxide stabilized/covered with chitosan polymer was formed on the metal surface, which protected the metal from further corrosion that contributed to the increased R_{ct} values at 24 h immersion. At the same time, double layer capacitance remained constant indicating the formation of continuous oxide layer on the metal surface. The deposition of iron oxide covered with chitosan backbone effectively blocked the damages on the chitosan coating that might have contributed to an increased R_p values during the initial stages of immersion in salt solution. This passive oxide layer was clearly observed in the optical photograph of PVB_Ch_PVB coating taken after 24 h EIS and film removal (Figure 7.4c). Raman spectra of this film showed new bands at 407 cm^{-1}, 502 cm^{-1}, 605 cm^{-1} and 656 cm^{-1} attributed to Fe_3O_4 oxides and characteristic Raman shifts of γ- Fe_2O_3 at 224 cm^{-1}, 291 cm^{-1} and 1310 cm^{-1} [38,39].

7.3.4 Effect of Chitosan Crosslinking with Glutaraldehyde on the Protective Behavior of lbl Coatings

Figure 7.6a shows EIS spectra acquired after 2 h immersion in 0.3 M salt solution for PVB_Ch_PVB coatings, where chitosan layer was crosslinked by the incorporation of different percentages of

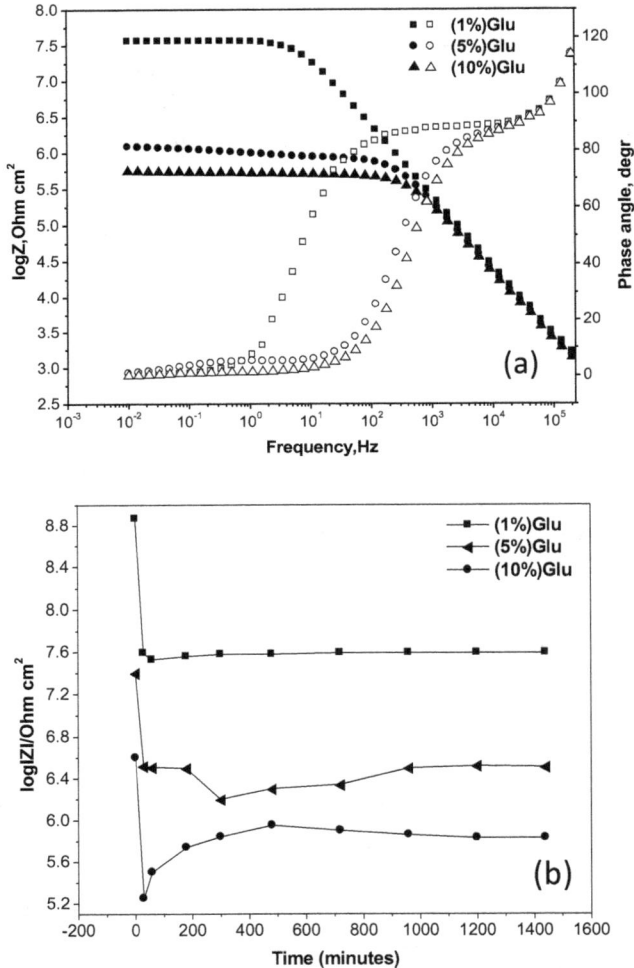

Figure 7.6 EIS plots of PVB_Ch/x%Glu_PVB coating having different percentages of glutaraldehyde; (a) Bode and phase plots obtained after 2 h immersion in 0.3 M salt solution and (b) logZ at low frequency from the Bode plot vs. time of immersion.

glutaraldehyde. For all the crosslinked coatings, Z-modulus at low frequency was higher than un-crosslinked PVB_Ch_PVB coating. 1% glutaraldehyde in the chitosan layer increased log Z to 7.6 Ω cm^2, though higher percentages of glutaraldehyde (10%) decreased log Z to 5.7 Ω cm^2. However, it was still higher than log Z value of 4.9 Ω cm^2 observed for un-crosslinked PVB_Ch_PVB coating. The high value of low frequency impedance showed the enhanced corrosion resistance property of PVB_Ch/x%Glu_PVB coatings, which might be due to the formation of crosslinked structure as observed in the IR spectra (Figure 7.1) which restricted the penetration of corrosive species to the metal surface by acting as a physical barrier. The barrier response of the coatings was also obvious in the phase plots (in Figure 7.6a) with a time constant at high frequency region (10^5 Hz). The identical slope in this frequency region indicated that the three coatings had identical thickness [40]. The phase angle remained close to 90° for half of the frequency range (10^2-10^5 Hz) covered in the measurement for PVB_Ch/1%Glu_PVB coating indicating a better barrier property than the PVB_Ch/5%Glu_PVB and PVB_Ch/10%Glu_PVB coatings. Second time constant corresponded to corrosion process on metal surface were apparent at 10^0 Hz for 5% and 10%Glu modified lbl coatings whereas for 1%Glu modified coating, it appeared as a small hump at 10^{-1} Hz. Nyquist plots in Figure 7.7 also demonstrate these changes in detail.

Nyquist plots of PVB_Ch/x%Glu_PVB coatings were fitted with the electrical circuit presented in Figure 7.5 and the fitting results are given in Table 7.2. Pore resistance (R_p) of the PVB_Ch_PVB coating increased significantly by the addition of 1% glutaraldehyde in the chitosan layer. Coating capacitance was observed to decrease, which indicated that the PVB_Ch/1%Glu_PVB coating was less permeable to water. However, the capacitance of the coatings increased with increasing glutaraldehyde dosage. These changes can be explained as follows: as the crosslinking agent (glutaraldehyde) content was low, the crosslinking degree was also low, and the molecular chains were in a coiled state which did not form a fixed passage, thus reducing the water absorption. When the glutaraldehyde content in the coating increased, the crosslinking degree of the coating also increased, and the crosslinked networks generated a basic structure which could form a fixed passage for the water molecules [41]. Water/ion transport through the coatings reduced the pore resistance which was evident from the decreasing R_p values of PVB_Ch/5%Glu_PVB and PVB_Ch/10%Glu_PVB coatings. When the time of immersion was increased, coating capacitance (C_c) and pore resistance (R_p) of PVB_Ch/1%Glu_PVB coatings increased initially followed by constant

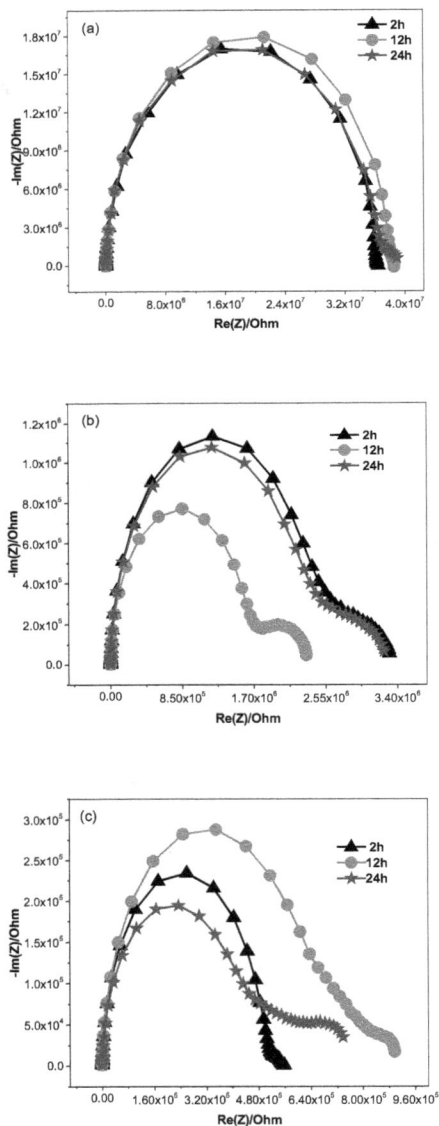

Figure 7.7 Nyquist plots of (a) PVB_Ch/1%Glu_PVB coating, (b) PVB_Ch/5%Glu_PVB coating and (c) PVB_Ch/10%Glu_PVB coating at different times of immersion in 0.3 M salt solution.

C_c for over 24 h whereas R_p decreased slightly. In addition, after an initial decrease, low frequency impedance remained almost constant over the measurement time (Figure 7.6b). These results indicated an enhanced protective behavior of PVB_Ch/1%Glu_PVB coating over the neat PVB_Ch_PVB coating. Crosslinking of chitosan with glutaraldehyde reduced the extent of hydrogen bonding between chitosan chains, demonstrated earlier through IR spectra, and released more free hydroxyl and amine groups suitable for the chelation of iron ions that led to the formation of stable passive oxide layer on the

Table 7.2 EIS parameters of PVB_Ch/x%Glu_PVB coatings at different time of immersion in 0.3 M salt solution

Coatings	Exposure time (h)	R_p (Ωcm^2)	C_c (Fcm^2)	Q_c	
				Y_0 (F.s^.n-1)	n
PVB_Ch/ 1%Glu_ PVB	2	36.12e6	8.84e-5	0.75e-9	0.97
	12	38.81e6	9.57e-5	0.81e-9	0.97
	24	37.57e6	9.60e-5	0.82e-9	0.97
PVB_Ch/ 5%Glu_ PVB	2	0.84e6	0.98e-4	0.97e-9	0.96
	12	2.235e6	1.06e-4	0.99e-9	0.96
	24	2.389e6	1.05e-4	0.99e-9	0.96
PVB_Ch/ 10%Glu_ PVB	2	0.50e6	1.07e-4	1.08e-9	0.96
	12	0.56e6	1.09e-4	1.13e-9	0.96
	24	0.36e6	1.10e-4	1.17e-9	0.95

Coatings	Exposure Time (h)	R_{ct} (Ωcm^2)	C_{dl} (Fcm^2)	Q_{dl}	
				Y_0 (F.s^.n-1)	n
PVB_Ch/ 1%Glu_ PVB	2				
	12	-	-	-	-
	24				
PVB_Ch/ 5%Glu_ PVB	2	0.38e6	4.92e-3	1.25e-6	0.58
	12	1.06e6	2.41e-3	0.41e-6	0.53
	24	0.89e6	3.05e-3	0.49e-6	0.52
PVB_Ch/ 10%Glu_ PVB	2	6.33e4	3.60e-2	9.85e-6	0.43
	12	20.41e4	6.30e-3	1.11e-6	0.44
	24	46.36e4	1.10e-2	1.21e-6	0.30

metal surface. π electrons in the double bonds of azomethine groups formed by the reaction of chitosan amine with glutaraldehyde carbonyl could also act as an active site of iron oxide stabilization process. Such a corrosion protection was obvious in the digital image of steel surface after removal of PVB_Ch/1%Glu_PVB coating given in Figure 7.8a that showed a uniformly covered grayish layer on the surface. Wessling [38] and Lu *et al.* [39] reported grayish colored layer

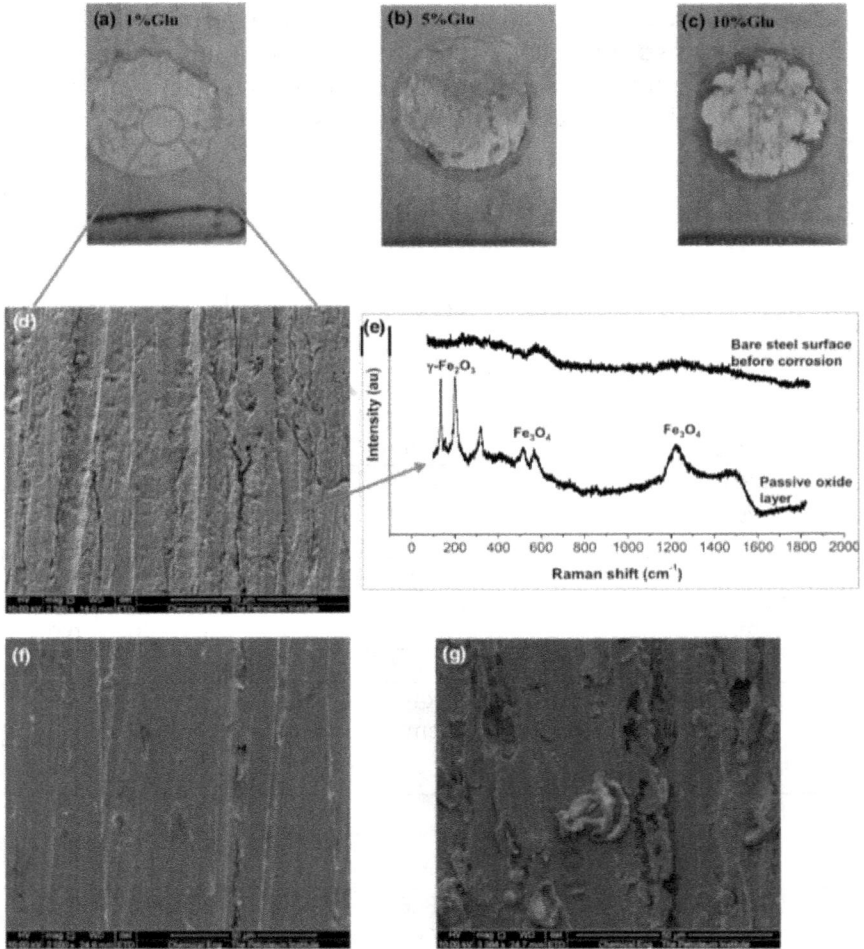

Figure 7.8 Optical photographs of (a) PVB_Ch/1%Glu_PVB, (b) PVB_Ch/5%Glu_PVB and (c) PVB_Ch/10%Glu_PVB coated steel surface after 24 h EIS in 0.3 M salt solution and film removal. Diameter of the circle is 1 cm². (D) SEM image and (E) Raman spectra of the area marked with red circle on optical photograph of steel surface taken after PVB_Ch/1%Glu_PVB coating removal. SEM images of bare steel (F) before and (G) after 24 h EIS measurement in 0.3 M salt solution.

as resulting from the formation of passive oxide layer on metal surface. SEM images and Raman spectra of these samples were analyzed to study the changes on the metal surface. The white lines or scratches observed in the SEM micrograph of bare steel in Figure 7.8f, taken before corrosion analysis, were formed by the sand paper polishing process. Raman spectra of this area showed a small hump of iron oxide at 600 cm^{-1} (Figure 7.8e), which indicated that acid itching and sand paper polishing removed iron oxide impurities to an extent. Porous corrosion products formed as clusters on the bare steel surface while immersion in salt solution are obvious in Figure 7.8g, whereas a less porous uniform film was observed in the SEM image of steel surface after 24 h EIS and removal of PVB_Ch/1%Glu_PVB coating (Figure 7.8d). Raman spectra (Figure 7.8e) of this film showed characteristic bands of Fe_3O_4 oxides at 408 cm^{-1}, 498 cm^{-1}, 606 cm^{-1} and 659 cm^{-1}, and characteristic Raman shifts of γ-Fe_2O_3 at 225 cm^{-1} and 290 cm^{-1} [38,39]. It can be assumed from these results that infinite array of functional groups on chitosan could act as a chain of nucleation sites from which further growth of iron oxide phase took place. Ultimately, this process resulted in a uniform layer of iron oxide covered with chitosan backbone on the metal surface. In the absence of complexing agent, these iron oxide particles agglomerate and form macroscopic precipitate which was observed on the bare steel surface as clusters of corrosion products. Chitosan steric stabilization prevented agglomeration of iron oxide particles which resulted in a uniformly distributed passive layer on the metal surface beneath the lbl coatings [31].

For PVB_Ch/5%Glu_PVB coating, low frequency impedance (logZ), given in Figure 7.6b, initially decreased, followed by an increase and subsequently remaining constant over the measurement time indicating corrosion protection of the coating, though it had a low logZ modulus as compared with PVB_Ch/1%Glu_PVB coating. It was evident from the C_c and R_p values in Table 7.2, where the C_c values remained almost constant and R_p increased with immersion time, which indicated that coating prevented the localization of water at the coating/metal interface as free electrolyte. However, high double layer capacitance (C_{dl}) and low charge transfer resistance (R_{ct}) values of PVB_Ch/5%Glu_PVB coatings after 24 h immersion indicating initiation of active corrosion processes on the metal surface, which are obvious in the optical photographs taken after film removal (Figure 7.8b) that showed corrosion products and black coloration. Thick corrosion product was visualized on the metal surface after EIS measurement and removal of PVB_Ch/10%Glu_PVB coating (Figure 7.8c), which concluded that the coating did not provide any corrosion protection. It was revealed from the R_p values as well, which decreased with time of immersion. At the same time, C_c showed an increasing trend, as expected. Significantly high R_{ct} values observed at the end of the measurement, obtained by the protection provided by the thick corrosion products, evidenced the corrosion process on the PVB_Ch/10%Glu_PVB coated metal surface. The protective behavior of crosslinked chitosan coatings was further evaluated from the Tafel plots displayed in Figure 7.9a. For PVB_Ch/1%Glu_PVB coating, a substantial reduction in both anodic and cathodic current densities was observed. The suppression of the cathodic process was attributed to the barrier property of the coating to the diffusion of water and oxygen. The passive oxide layer stabilized by chitosan polymer backbone suppressed the dissolution of metal from the substrate that resulted in an inhibition of anodic process and a shift of the corrosion potential (E_{corr}) to the positive direction [36]. Further addition of glutaraldehyde shifted E_{corr} to more anodic side supporting the aforementioned EIS data.

Figure 7.9a also demonstrates Tafel plots of PVB_Ch/1%Glu_PVB coatings with graphene and vermiculite incorporated in chitosan layer. Bode and phase plots of these coatings taken after 2 h immersion in 0.3 M salt solution are shown in Figure 7.9b. Compared with PVB_Ch/1%Glu_PVB coating, low frequency impedance (logZ) decreased to an order of nearly one and three respectively for PVB_Ch/1%Glu/5%Gr_PVB and PVB_Ch/1%Glu/5%Ver_PVB coatings which indicated that incorporation of graphene and vermiculite reduced the corrosion resistance of PVB_Ch/1%Glu_PVB coating. It was earlier reported that well dispersed inorganic particles could increase the length of the diffusion pathways for the gas and vapor molecules in the polymer coatings that could lead to a significant enhancement of corrosion protection of metallic substrates as compared with that of a neat polymer coating [42]. However, in the present case, the reduced corrosion protection might have resulted from the poor barrier properties due to imperfect adhesion of the fillers to the polymer matrix, which caused the formation of additional diffusion pathways [34]. No interaction between the fillers and the polymer matrix was detected in the IR spectra of these coatings (Figure 7.9c) as the characteristic vibrations of PVB_Ch/1%Glu_PVB coating didn't exhibit any change with the incorporation of graphene and vermiculite in the chitosan layer. Poor barrier performance was evident

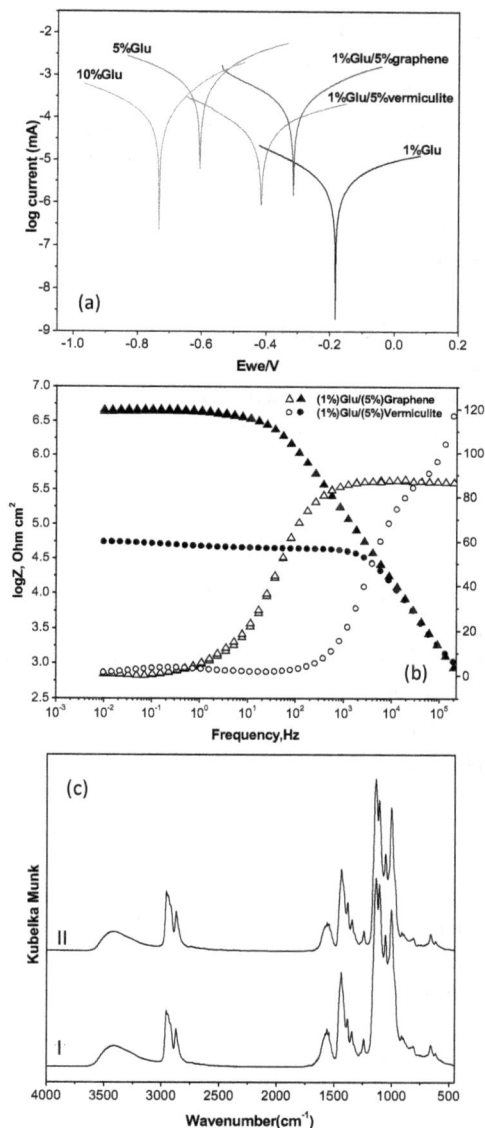

Figure 7.9 (a) Tafel plots of PVB_Ch/1%Glu_PVB, PVB_Ch/5%Glu_PVB and PVB_Ch/10%Glu_PVB, PVB_Ch/1%Glu/5%Gr_PVB and PVB_Ch/1%Glu/5%Ver_PVB coatings recorded after 7 h immersion in 0.3 M salt solution, (b) Bode and phase plots of PVB_Ch/1%Glu/5%Gr_PVB and PVB_Ch/1%Glu/5%Ver_PVB coatings obtained after 2 h immersion in 0.3 M salt solution and (c) IR spectra of PVB_Ch/1%Glu/5%Gr_PVB coating (I) and PVB_Ch/1%Glu/5%Ver_PVB coating (II).

from the phase angle plot at high frequency region as well. For PVB_Ch/1%Glu/5%Gr_PVB coating, the plateau in the high frequency region reduced slightly to the frequency range of 10^3-10^5 Hz. However, for PVB_Ch/1%Glu/5%Ver_PVB coating, the plateau in the high frequency region reduced significantly showing poorer barrier performance due to introduction of new diffusion pathways for the oxygen and water vapor in the polymer coating. Extensive polarity of vermiculite also makes it more susceptible to moisture uptake, thus, resulting in deterioration of anti-corrosion performance. The shifting of E_{corr} in the Tafel plots of graphene and vermiculite incorporated coatings to -0.32v and -0.40v respectively, thus, confirming further the lower extent of corrosion protection of these coatings compared with neat PVB_Ch/1%Glu_PVB coating (-0.18v).

7.4 Conclusions

Layer-by-layer addition of chitosan and PVB was employed to generate multi-layer coatings on the carbon steel substrate. Corrosion inhibition of chitosan coatings, brought about by the interaction of metal ions with its functional groups, was confirmed owing to the high Z-modulus at low frequency in the Bode plot. The reduced coating stability of neat chitosan, due to its high affinity towards water, could be successfully controlled by the sandwiching of chitosan layer between two layers of PVB coatings. Incorporation of glutaraldehyde in the chitosan layer in PVB_Ch_PVB coatings introduced a cross-linked structure, resulting in a physical barrier to water and ions reaching the metal substrate. High R_p and low C_c values of PVB_Ch/1%Glu_PVB coating at the initial time of immersion supported the improved coating behavior. Chitosan backbone was observed to be involved in the stabilization process of passive iron oxide layer on metal surface by iron ion's chelation with free amine and hydroxyl groups of chitosan followed by further oxidation. In other words, an infinite array of donor groups on the chitosan backbone acted as a chain of nucleation sites from which further growth of iron oxide layer took place. Finally, a passive oxide layer sterically stabilized with chitosan backbone was formed that was visually observed as a grayish layer covering the steel surface uniformly. SEM and Raman spectra of the oxide layer confirmed the formation of Fe_3O_4 and γ-Fe_2O_3 oxides on the metal surface. Extent of chitosan crosslinking was the determining factor for the stability and corrosion protection of PVB_Ch/x%Glu_PVB coatings. High glutaraldehyde content in the chitosan layer decreased coating performances as the extensive crosslinking provided a fixed path for the penetration of water and ions through the coating. It can be suggested from these results that layer-by-layer addition of chitosan and PVB would be a promising coating method to fabricate stable chitosan coatings. Furthermore, by controlling the extent of crosslinking between the chitosan chains, corrosion inhibition property as well as stability of the coatings can be efficiently improved.

Summary of the Results

Anti-corrosion coatings of chitosan were fabricated by layer-by-layer (lbl) addition of chitosan (Ch) and poly vinyl butyral (PVB) on mild carbon steel substrate. Corrosion inhibition of chitosan in lbl coatings was analyzed by electrochemical, spectroscopic and morphological measurements. Sandwiching of chitosan between two hydrophobic PVB layers enhanced its bonding strength and enabled the amine and hydroxyl groups of chitosan to chelate with iron ions that led to the formation of a chitosan stabilized iron oxide passive layer on the metal surface. Effect of chitosan crosslinking on the stabilization process of passive oxide layer as well as the coating stability was studied by incorporating glutaraldehyde in the chitosan layer. The results obtained from electrochemical impedance spectroscopy and Tafel plots clearly showed a superior corrosion protection of lbl coatings generated by 1% glutaraldehyde incorporation in the middle chitosan layer. SEM and Raman spectra confirmed the formation of passive oxide layer on the metal surface stabilized by chitosan polymer backbone. Improvement in the anti-corrosion properties of PVB_Ch/1%Glu_PVB coatings were not observed after the incorporation of fillers like graphene and vermiculite in the chitosan layer probably due to the imperfect adhesion of the fillers with the polymeric matrix.

Acknowledgements

The authors sincerely acknowledge the financial support from Petroleum Institute Gas Processing and Materials Science Research Center (GRC) for the project GRC-005.

This work has been published earlier in Cellulose (2015) 22:3275-3290. The work has been reproduced here with permission from Springer.

References

1. Gonzalez-Garcia, Y., Gonzalez, S., and Souto, R. M. (2007) Electro-chemical and structural properties of a polyurethane coating on steel substrates for corrosion protection. *Corrosion Science*, **49**, 3514-3526.
2. Leidheiser, H. (1982) Corrosion of painted metals – A review. *Corrosion*, **38**, 374-383.
3. Walter, G. W. (1986) A critical review of the protection of metals by paints. *Corrosion Science*, **26**, 27-38.
4. Gandini, A., and Belgacem, M. N. (2002) Recent contribution to the preparation of polymers derived from renewable resources. *Journal of Polymers and the Environment*, **10**, 105-114.
5. Derksen, J. T. P., Cuperus, F. P., and Kolster, P. (1996) Renewable re-sources in coatings technology: A review. *Progress in Organic Coatings*, **27**, 45-53.
6. Carneiro, J., Tedim, J., Fernandes, S. C. M., Freire, C. S. R., Silvestre, A. J. D., Gandini, A., Ferreira, M. G. S., and Zheludkevich, M. L. (2012) Chitosan-based self-healing protective coatings doped with cerium nitrate for corrosion protection of aluminium alloy 2024. *Progress in Organic Coatings*, **75**, 8-13.
7. Pillai, C. K. S., Paul, W., and Sharma, C. P. (2009) Chitin and chitosan polymers, chemistry, solubility and fiber formation. *Progress in Polymer Science*, **34**, 641-678.
8. Kumar, G., and Buchheit, R. G. (2006) Development and characteri-zation of corrosion resistant coatings using the natural biopolymer chitosan. *ECS Transactions*, **1**, 101-117.
9. Lundvall, O., Gulppi, M., Paez, M. A., Gonzaleza, E., Zagal, J. H., Pavez, J., and Thompson, G. E. (2007) Copper modified chitosan for protec-tion of AA-2024. *Surface and Coatings Technology*, **201**, 5973-5978.
10. Pang, X., and Zhitomirsky, I. (2007) Electrophoretic deposition of composite hydroxyapatite-chitosan coatings. *Materials Characteri-zation*, **58**, 339-348.
11. El-Sawy, S. M., Abu-Ayana, Y. M., and Abdel-Mohdy, F. A. (2001) Some chitin/chitosan derivatives for corrosion protection and waste water treatments. *Anti-corrosion Methods and Materials*, **48**, 227-234.
12. Sugama, T., and Cook, M. (2000) Poly(itaconic acid)-modified chi-tosan coatings for mitigating corrosion of aluminum substrates. *Progress in Organic Coatings*, **38**, 79-87.
13. Ahmed, R. A., Farghali, R. A., and Fekry, A. M. (2012) Study for the stability and corrosion inhibition of electrophoretic deposited chi-tosan on mild steel alloy in acidic medium. *International Journal of Electrochemical Science*, **7**, 7270-7282.
14. Carneiro, J., Tedim, J., Fernades, S. C. M., Freire, C. S. R., Gandini, A.,

Ferreira, M. G. S., and Zheludkevich, M. L. (2013) Chitosan as a smart coating for controlled release of corrosion inhibitor 2-mercapto-benzothiazole. *ECS Electrochemistry Letters*, **2**, C19-C22.

15. Zheludkevich, M. L., Tedim, J., Freire, C. S. R., Fernades, S. C. M., Kallip, S., Lisenkov, A., Gandini, A., and Ferreira, M. G. S. (2011) Self-healing protective coatings with "green" chitosan based pre-layer reservoir of corrosion inhibitor. *Journal of Materials Chemistry*, **21**, 4805-4812.

16. Bumgardner, J. D., Wiser, R., Gerard, P. D., Bergin, P., Chesnutt, B., Marin, M., Ramsey, V., Elder, S. H., and Gilbert, J. A. (2003) Chitosan, potential use as a bioactive coating for orthopaedic and craniofacial/dental implants. *Journal of Biomaterials Science, Polymer Edition*, **14**, 423-438.

17. Sugama, T., and Jimenez, S. M. (1999) Dextrine-modified chitosan marine polymer coatings. *Journal of Materials Science*, **34**, 2003-2014.

18. El-Haddad, M. N. (2014) Hydroxyethylcellulose used as an eco-friendly inhibitor for 1018 c-steel corrosion in 3.5% NaCl solution. *Carbohydrate Polymers*, **112**, 595-602.

19. Umoren, S. A., Banera, M. J., Garcia, T. A., Gervasi, C. A., and Mirifico, M. V. (2013) Inhibition of mild steel corrosion in HCl solution using chitosan. *Cellulose*, **20**, 2529-2545.

20. Manlin, L., Juan, X., Ronghua, L., Dongen, W., Tianbao, L., Maosen, Y., and Jinyi, W. (2014) Simple preparation of aminothiourea-modified chitosan as corrosion inhibitor and heavy metal ion adsorbent. *Journal of Colloid and Interface Science*, **417**, 131-136.

21. Mohammed, R. R., and Fekry, A. M. (2011) Antimicrobial and anti-corrosive activity of adsorbents based on chitosan Schiff's base. *International Journal of Electrochemical Science*, **6**, 2488-2508.

22. Zheludkevich, M. L., Serra, R., Montemor, M. F., Yasakau, K. A., Miranda Salvado, I. M., and Ferreira, M. G. S. (2005) Nanostructured sol–gel coatings doped with cerium nitrate as pre-treatments for AA2024-T3, corrosion protection performance. *Electrochimica Acta*, **51**, 208-217.

23. Hsu, C. S., and Mansfed, F. (2011) Concerning the conversion of the constant phase element parameter Yo into a capacitance. *Corrosion*, **57**, 747-748.

24. Lian, F., Wen, Y., Ren, Y., and Guan, H. Y. (2014) A novel PVB based polymer membrane and its application in gel polymer electrolytes for lithium-ion batteries. *Journal of Membrane Science*, **456**, 42-48.

25. Tripathy, A. R., Chen, W., Kukureka, S. N., and MacKnight, W. J. (2003) Novel poly(butylene terephthalate)/poly(vinyl butyral) blends prepared by in-situ polymerizations of cyclic poly(butylene terephthalate) oligomers. *Polymer*, **44**, 1835-1842.

26. Li, B., Shan, C. L., Zhou, Q., Fang, Y., Wang, Y. L., Xu, F., Han, L. R.,

Ibrahim, M., Guo, L. B., Xie, G. L., and Sun, G. C. (2013) Synthesis, characterization, and antibacterial activity of cross-linked chitosan-glutaraldehyde. *Marine Drugs*, **11**, 1534-1552.

27. Sipos, P., Berkesi, O., Tombacz, E., St. Pierre, T. G., and Webb, J. (2003) Formation of spherical iron(III) oxyhydroxide nanoparticles sterically stabilized by chitosan in aqueous solutions. *Journal of Inorganic Biochemistry*, **95**, 55-63.

28. Hernandez, R., Zamora-Mora, V., Sibaja-Ballestero, M., Vega-Baudrit, J., Lopez, D., and Mijangos, C. (2009) Influence of iron oxide nanoparticles on the rheological properties of hybrid chitosan ferrogels. *Journal of Colloid and Interface Science*, **339**, 53-59.

29. Oyrton, A. C., Monteiro, J., and Claudio, A. (1999) Some studies of crosslinking chitosan-glutaraldehyde interaction in a homogeneous system. *International Journal of Biological Macromolecules*, **26**, 119-128.

30. Socrates, G. (1994) *Infrared and Raman Characteristics Group Frequencies*, 3rd Edition, John Wiley & Sons Ltd., USA.

31. Qian, M., Soutar, A. M., Tan, X. H., Zeng, X. T., and Wijesinghe, S. L. (2009) Two-part epoxy-siloxane hybrid corrosion protection coatings for carbon steel. *Thin Solid Films*, **517**, 5237-5242.

32. Eduok, U. M., and Khaled, M. M. (2014) Corrosion protection of steel sheets by chitosan from shrimp shells at acid pH. *Cellulose*, **21**, 3139-3143.

33. Bao, Q., Zhang, D., and Wan, Y. (2011) 2-Mercaptobenzothiazole doped chitosan/11-alkanethiolate acid composite coating, Dual function for copper protection. *Applied Surface Science*, **257**, 10529-10534.

34. Latnikova, A., Grigoriev, D., Schenderlein, M., Mohwald, H., and Shchukin, D. (2012) A New approach towards "active" self-healing coatings, exploitation of microgels. *Soft Matter*, **8**, 10837-10844.

35. Rout, T. K., Bandyopadhyay, N., and Venugopalan, T. (2006) Polyphosphate coated steel sheet for superior corrosion resistance. *Surface and Coatings Technology*, **201**, 1022-1030.

36. Radhakrishnan, S., Siju, C. R., Mahanta, D., Patil, S., and Madras, G. (2009) Conducting polyaniline-nano-TiO2 composites for smart corrosion resistant coatings. *Electrochemica Acta*, **54**, 1249-1254.

37. Colreavy, J., and Scantlebury, J. D. (1995) Electrochemical impedance spectroscopy to monitor the influence of surface preparation on the corrosion characteristics of mild steel MAG welds. *Journal of materials processing technology*, **55**, 206-212.

38. Wessling, B. (1994) Passivation of metals by coating with polyaniline, corrosion potential shift and morphological changes. *Advanced Materials*, **6**, 226-228.

39. Lu, W. K., Elsenbaumer, R. L., and Wessling, B. (1995) Corrosion protection of mild steel by coatings containing polyaniline. *Synthetic*

Metals, **71**, 2163-2166.

40. Maia, F., Tedim, J., Lisenkov, A. D., Salak, A. N., Zheludkevich, M. L., and Ferreira, M. G. S. (2012) Silica nanocontainers for active corrosion protection. *Nanoscale*, **4**, 1287-1298.

41. Yu, Q., Song, Y., Shi, X., Xu, C., and Bin, Y. (2011) Preparation and properties of chitosan derivative/poly(vinyl alcohol) blend film crosslinked with glutaraldehyde. *Carbohydrate Polymers*, **84**, 465-470.

42. Chang, C. H., Huang, T. C., Peng, C. W., Yeh, T. C., Lu, H,, Hung, W., Weng, C. J., Yang, T., and Yeh, J. M. (2012) Novel anticorrosion coatings prepared from polyaniline/graphene composites. *Carbon*, **50**, 5044-5051.

8

Antifouling and Antibacterial Polymeric Membranes

Shabnam Pathan and Suryasarathi Bose*

Department of Materials Engineering, Indian Institute of Science, Bangalore 560012, India

*Corresponding author: sbose@iisc.ac.in

8.1 Introduction

Water is one of the basic necessities required for the sustenance and continuation of life, agriculture and industrial processes. It is, therefore, important that supply of good quality water should be available for various needs. The main reason for the water crisis is the rapid growth of population, urbanization and unsustainable economic practices, which result in large scale pollution. Hence, in order to have abundant water supply for human needs, it is necessary to find cost effective methods to use seawater and to reuse the contaminated water [1,2].

Contamination of water by various pathogens is one of the main problems faced on a regular basis. As per World Health Organization (WHO), millions of people die because of waterborne infections. Several methods have been adopted for disinfection like chlorination, ozonation, UV radiation, etc. However, these methods have shortcomings like regrowth of pathogens, formation of toxic disinfections by products and resistance to UV radiation. Further, the type of bacteria also influences the efficiency of the disinfection methods. The gram negative bacteria are more resistant in comparison to the gram positive bacteria because of the presence of extra protective layer. Henceforth, there is an urgent need of alternative approaches which are more effective in eliminating bacteria [3-7]. Various antibacterial agents have been used for eliminating different pathogens like silver nanoparticles, graphene oxide, quaternary ammonium salts, etc. However, these antibacterial agents also suffer from high capital cost, toxicity, efficiency and leaching tendency.

Marine Coatings and Membranes, edited by Vikas Mittal
© 2019 Central West Publishing, Australia

Number of integrated technologies like air flotation, flocculation, adsorption, distillation and advanced oxidization processes (AOPs) are available for seawater treatment as well as remediation of contaminated water. However, the above mentioned technologies use large amount of chemicals, produce toxic sludge, consume large amount of energy and are costly to operate. Henceforth, the development of sustainable, mechanically robust, environmentally friendly and low-cost technologies for water treatment is of supreme significance [8,9].

Recently, membrane technology has gained considerable interest because of simple, environmentally safe and scalable process for seawater treatment and contaminated water purification, as compared to conventional methods [10,11]. The polymeric membrane technology has several advantages like less chemical usage and sludge generation as well as high-quality permeates. Different types of polymeric marine membranes based on microfiltration (MF), ultrafiltration (UF), nanofiltration (NF), reverse osmosis (RO) and forward osmosis (FO) have been widely used. The advantages of polymeric membranes include: highly selective separation, easily scalable process, use of no/or little chemicals, cost effectiveness, low energy consumption, etc. [9]. However, fouling of membranes with time is one of the main drawbacks associated with membrane technology. Membrane fouling is the deposition of colloidal particles, organic, inorganic salts, etc., at the surface or in the membrane pores. Membrane fouling causes pore blocking, cake formation, organic adsorption, inorganic precipitation and biofouling, which leads to the declination of flux permanently or temporarily [12,13]. Membrane cleaning by chemicals has been widely used to regenerate fouled membranes. However, regular use of chemicals shortens the lifetime of polymeric membranes. Further, membranes need to be replaced once the chemical treatment becomes ineffective. Hence, membrane fouling is one of the most significant economic challenges faced by membrane technology [14]. As mentioned earlier, various types of foulants include inorganic, organic, and bio-foulants. Inorganic foulants include silica, aluminum silicate minerals and ferric oxide/hydroxide colloids. In contrast, organic foulants include oils, BSA, humic acid, etc., while biofoulants comprise many types of microorganisms including bacterial cells, algae, etc. Several strategies have been used to improve antifouling properties of membranes like physical blending, surface coating, surface grafting, inclusion of inorganic additives, surface bioadhesion, etc. In order to render antifouling properties to the membranes, chemical, physical and topological modifications have been proposed [9]. Additionally, another challenge in membrane technology is to fabricate antibacterial surfaces without compromising the overall operating cost. Hence, the primary focus should be high flux, good antibacterial characteristics, stability and mechanical strength [7]. This chapter comprehensively discusses polymeric marine membranes for water (seawater and contaminated water) remediation in general and strategies to impart antifouling and antibacterial properties in particular.

8.2 Membrane Fouling

Membrane fouling is an overwhelming challenge in membrane technology used for water remediation. It is a very serious problem in pressure-driven processes such as MF, UF, NF and RO. In pressure-driven membranes, membrane hydraulic resistance, concentration polarization resistance, cake layer resistance and adsorption resistance are the main factors responsible for flux declination [15-17].

Membrane fouling can be divided into reversible fouling and irreversible fouling. Reversible fouling can be removed after rinsing, however, irreversible fouling cannot be removed by physical rinsing because of the interactions between the foulants and membrane surface. The interactions can be of specific and non-specific nature. The specific interactions include covalent bonding/coordination interactions formed between the membrane and foulant, while van der Waals' interactions, hydrophobic interactions as well as hydrogen bonding between the foulant molecules and membrane surface are responsible for non-specific interactions. Organic and biofoulants are further classified into three categories: spreadable foulants, non-migratory foulants and biofoulants/proliferative foulants. Spreadable foulants get spread and form continuous layer on the membrane. Non-migratory foulants do not spread on the surface, however, form a stable cake layer. Proliferative foulants includes bacterial cells, extracellular polymeric substances and cell debris [18-20]. Proliferative fouling can occur through various steps like physical adsorption, electrostatic interactions, chemical bonding, mechanical interlocking and proliferation, etc. [21,22]. Type of polymers used in membrane fabrication has also an impact on fouling behavior. For instance, hydrophobic polymers, high surface roughness and less surface charge on the membranes are more prone to foul [9].

8.3 Strategies for Rendering Antifouling Surfaces

The various approaches used for improving antifouling properties are surface grafting, surface coatings, surface aggregation and surface bioadhesion, and the mechanisms involved in enhancing the antifouling properties using these approaches are mainly active and passive mechanisms [9]. In the following sections, we will discuss the various strategies adopted for enhancing the antifouling properties of the polymeric membranes.

8.3.1 Surface Coatings

Surface coating is one of the strategies for improving the antifouling properties of polymeric membranes. In this process, the membranes are either post modified with antifouling materials like poly(vinyl alcohol) (PVA)- and polyethylene glycol (PEG)-based polymers or inorganic nanomaterials are coated on the surface of the membranes via dip or spin coating [15]. Thin-gel composite UF membranes were prepared by Li and Barbari [23] via spin coating PVA hydrogels onto cellulose membranes. The PVA hydrogel coatings over UF membranes greatly reduced the irreversible protein fouling.

Polydopamine (PDA) has been widely used for imparting the antibacterial and antifouling properties to the polymeric membranes. For instance, Jiang *et al.* [24] used PDA for coating the hydrophobic porous polypropylene membranes. Later, the PDA coated membranes were dipped in PVP solution which got adhered via hydrogen bonding. The coated membranes showed high antifouling property and high flux rate, in addition to the antibacterial property through complexation of PVP layer by iodine complexation [24]. The overall scheme of the PDA coating as well as subsequent complexation of iodine and PVP is shown in Figure 8.1. In another study, Freeman *et al.* [25] fabricated UF and RO membranes coated with PEG-based graft polymers. The resulting membranes showed low fouling properties against oil/water emulsion.

For coating inorganic nanomaterials, the membrane surface is usually pre-treated to induce adsorption of inorganic precursors. Several techniques are used to facilitate the process *in-situ*, such as sol-gel, chemical reduction and biomineralization [26,27].

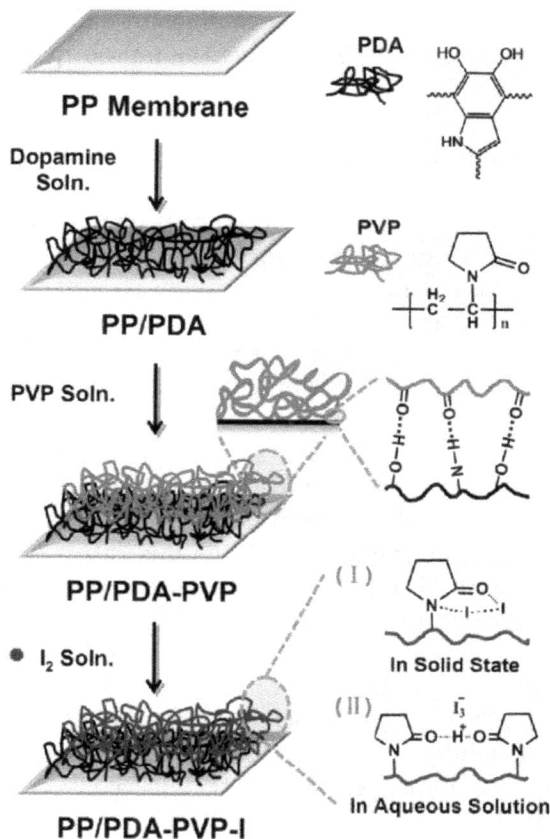

Figure 8.1 Schematic of PDA coating as well as subsequent complexation of iodine and PVP. Reproduced from Reference 24 with permission from American Chemical Society.

The chemical vapor deposition (CVD) technique was used to fabricate the antifouling membranes via free-radical polymerization technique. Gleason and co-workers incorporated zwitterionic coating onto RO membranes surface via CVD technique [28-32]. For this purpose, a series of antifouling copolymers such as poly(hydroxyethylmethacrylate-coperfluorodecylacrylate) (P(HEMA-co-PFDA)), P4VP and its copolymers were coated onto RO membrane surface. Although surface coating can be successfully applied for antifouling functionality with high coverage density, however, it gets detached during long term operation due to weak non-covalent interactions with the polymeric membrane. Furthermore, by surface coating strategies, some changes in membrane pore size may also occur which may also change the membrane permeability and selectivity [28-32].

8.3.2 Polymer Blending

Blending of hydrophilic polymeric materials or polymers with inorganic compounds is one of the approaches for improving the antifouling property of the marine membranes. The main advantage of this approach is that it can be easily adapted to existing membrane fabrication processes [33]. By using this approach, UF membranes with enhanced protein adsorption-resistant ability were prepared by Wang *et al.* [34]. PVP and PEG have been widely used as additives in the preparation of poly(ether sulfone) (PES) based UF membranes [35,36]. Molar mass of PEG has also been observed to influence the resulting membrane performance. The membranes fabricated using higher molar mass have high flux rate and larger pores. Liu *et al.* [37] found an optimum PEG content in the PES-NMP (N-methylpyrrolidone) system for increasing water permeability. The study revealed that the flux rate increased and BSA adsorption decreased [37]. Mural *et al.* [38] used PEG as one of the components in polyethylene-grafted GO (PE-g-GO) and its combination with maleated PE (maleic anhydride-grafted PE) for enhancing the antifouling properties of PE membranes [38]. A triblock copolymer PEO-PPO-PEO, commonly known as Pluronic, was used for the modification of UF membranes [39]. The presence of PEO chain produced repulsion effect and consequently prevented the adsorption of protein on the membrane surface.

It has been reported in the literature that amphiphilic polymers with different structures can be used to improve the antifouling property of UF membranes. The low biofouling of membranes modified by amphiphilic or surface active additives is because of the heterogeneities in the structure as well as composition of the surface.

However, some drawbacks are still associated with blending method, such as the poor compatibility between the organic polymer and inorganic nanomaterials, which may lead to some defects while fabricating the membranes. Secondly, in this method, most of the inorganic nanomaterials are mainly buried within the bulk polymer and not available on the membrane surface, thus, reducing the efficiency of the membrane as the antifouling property is a surface phenomenon [33,40,41].

8.3.3 Functionalization by Grafting-from and Grafting-to Membrane Surfaces

Functionalization by grafting-from and grafting-to membrane surfaces has been extensively used to control the fouling of the polymeric membranes. The effects of surface charge, surface roughness and surface hydrophilicity on bacteria attachment on chemically modified polyethylene membranes were systematically studied [42]. The study revealed that the hydrophilic, non-charged and smooth surfaces had the lowest biofilm coverage as compared to the unmodified polyethylene membrane. Zhang *et al.* [43] synthesized polyamide membranes grafted with a zwitterionic poly(sulfobetaine methacrylate) (pSBMA). The modified polyamide membranes had high flux rate with 97% reduction in protein adsorption. Similarly, Meng *et al.* [44] grafted zwitterionic poly(4-(2-sulfoethyl)-1-(4-vinylbenzyl) pyridinium betaine) (PSVBP) onto the surface of a polyamide membrane. The modified polyamide membrane showed good salt rejection and antifouling properties, but failed for long term operation. Li *et al.* [45] synthesized zwitterionic-catechol conjugates using grafting-to approach for enhancing anti-biofouling characteristic. Besides, plasma treatment is also used for modifying the membrane surface by introducing polar functional groups. Khulbe *et al.* [46] and van der Bruggen [47] used this approach for the modification of PES NF membranes, however, it can also be used for other polymeric membrane materials, thus, confirming the versatility of the approach.

Reversible addition-fragmentation chain transfer polymerization (RAFT) and atom-transfer radical polymerization (ATRP) techniques are used in grafting-from method [48]. For instance, Elimelech and co-workers synthesized functional block copolymer brushes on commercial thin-film composite (TFC) membranes by ATRP technique using mussel-inspired catechol chemistry, and the study demonstrated the integrated passive and active defense mechanism against biofouling [49].

In conclusion, grafting strategy reinforces the strong interaction between the grafted layer and membrane surface. Though, antifouling performance usually gets enhanced with increasing the grafting density, however, water permeability decrease in some cases with high grafting density. Hence, it is important to graft an optimum layer density on the membrane surface without compromising the water permeability of the membrane.

8.3.4 Inorganic Additives

The inclusion of inorganic nanoparticles like silica, Al_2O_3, titanium dioxide, ZrO_2 and Fe_3O_4 into polymeric membranes is also used for improving the antifouling properties [35]. It has been reported that the addition of nanoparticles significantly improves the membrane performance in terms of flux rate and antifouling properties.

The antifouling properties of the poly(vinylidene fluoride) (PVDF) membranes with and without TiO_2 nanoparticles were compared, and the study revealed that the TiO_2 nanoparticles with smaller size resulted in good antifouling performance [50]. Besides, the membranes fabricated using mixture of TiO_2 and SiO_2 nanoparticles resulted in high flux and low antifouling behavior [51]. Yu *et al.* [52] synthesized PVDF composite membranes using different wt% of silica nanoparticles. Despite the good reduction in fouling of membranes by adding TiO_2 nanoparticles, it was not possible to transfer this approach from the lab scale to the commercial level as the use of UV light source in the membrane module was not feasible at larger scale. Some of the methods used for incorporating the inorganic nanoparticles in polymeric membranes are mentioned below [35]:

1. Inclusion of inorganic nanoparticles directly in the casting solution via phase inversion.
2. Addition of the sol solution containing nanoparticles to the casting solution.
3. Dipping the membranes in the aqueous solution containing nanoparticles, and preparation of the composite membranes takes place via self-assembly.

8.4 Antibacterial Agents

As mentioned earlier, it has been demonstrated that the membranes with hydrophilic moieties and smooth surface are less prone to foul, in comparison to the hydrophobic and rough surface. However, these factors focus only on the bacterial attachment, not on the bacterial elimination and growth. Hence, in order to develop multifunctional membranes, the strategies that can render the membrane surface antibacterial potentially provide an additional layer of protection along with the antifouling surface. Membranes modified with biostatic/biocidal moieties can help in the elimination or growth prevention of bacteria. This strategy can also help in preventing the fouling of the polymeric membranes caused by bacterial growth. The following sections discuss the different methods used for inducing the antibacterial properties on the membranes surface.

8.4.1 Inorganic Materials based Antibacterial Agents

Nanoparticles of copper oxide (CuNPs) and its derivatives are well known antibacterial agents. Copper NPs damage proteins and DNA of bacterial cells by forming hydroxyl radicals [53,54]. However, leaching of Cu ions with time is one of the main problem. Hence, in order to reduce the leaching of CuNPs from polymer matrices, it is essential to immobilize the nanoparticles on the polymeric surface by various strategies. For instance, antibacterial membranes fabricated by immobilization of copper nanoparticles on poly(acrylonitrile) (PAN) membrane surface using polyethyleneimine (PEI) showed good flux with minimum leaching of copper ions. Besides, the modified membrane showed good antibacterial activity against E. coli bacteria [55]. Chen *et al.* [56] investigated the antibacterial performance of PES membranes modified with halloysite nanotubes/copper ions hybrids. The modified membranes were 100% bacteriostatic against both gram negative and gram positive bacteria. However, the leaching profile and long term bactericidal efficiency of the membranes were not discussed in the study.

Other metallic oxide NPs such as ZnO, TiO_2 and Al_2O_3 have also been studied for developing antibacterial membranes. For instance, zinc oxide nanoparticles (ZnONPs) are considered to be excellent antibacterial agent because of non-toxicity, biocompatibility, hydrophilicity and photocatalytic activity. The ZnONPs show antibacterial activity due to the release of Zn^{2+} ions or the formation of reactive oxygen species (ROs) photocatalytically [57,58]. Hinai *et al.* [59] compared the antibacterial properties of ZnO nano-rods and ZnO nanoparticles incorporated in PES membranes. The study revealed that ZnO nano-rods were highly effective in eliminating the E. coli bacteria, relative to ZnO nanoparticles, due to the high surface area of the former. In another study, PSF (polysulfone) membranes functionalized with ZnONPs showed outstanding antibacterial activity against E. coli. However, the leaching of nanoparticles was not discussed in the study. The antibacterial efficiency of ZnO and ZnO-GO incorporated in PSF membrane was compared and the study revealed that 0.6 wt% ZnO-GO (as compared to 2 wt% ZnO) nanoparticles inhibited the growth of E. coli bacteria due to the synergistic effect of ZnO and GO sheets, which improved the dispersion of ZnO in presence of GO sheets. Titanium dioxide nanoparticles have high biocidal activity because of their capability of generating the oxidative stress, which can irreversibly damage the cell membrane. Besides, due to highly hydrophilic nature, TiO_2 NPs reduce the fouling of the polymeric membranes. Mixed matrix membranes using PVDF and nano-sized TiO_2 particles were prepared by Damodar *et al.* [61], which revealed good antibacterial activity under UV irradiation. Another study reported the synthesis of mixed matrix membranes using PVDF, sulfonated polyethersulfone (SPES) and TiO_2 NPs (4wt%) with outstanding antibacterial efficiency [62]. Silver-impregnated TiO_2/nylon-6 nanocomposite mats were prepared by Pant *et al.* [63] with exceptionally high antibacterial and photocatalytic activity.

8.4.2 Carbon based Nanoparticles

Of late, carbon nanotubes (CNTs) have been widely used as antibacterial agents in polymeric membranes owing to excellent biocompatibility and contact elimination mechanism [64]. The incorporation of CNTs in polymeric membranes leads to mechanical and oxidative stress in bacteria. Further, it has been reported that single walled CNTs exhibit higher toxicity towards bacterial cells as compared to the multiwalled CNTs. Schiffman and Elimelech [65] reported superior antibacterial activity of electrospun polymer mats with incorporated single walled CNTs, as illustrated in the scanning electron micrographs (SEM) in Figure 8.2. The antibacterial action of the membranes was observed after a short contact period of 15 min or less. Besides, acid treated CNTs (with carboxyl groups) based membranes are considered to have higher antibacterial efficiency as compared to untreated CNTs. For instance, the PES membrane modified with functionalized CNTs exhibited good antibacterial properties [66].

Graphene oxide (GO) based nanomaterials, with one-atom thick laminar structure, have gained considerable interest as antibacterial agents due to the presence of polar functional groups like hydroxyl, carboxyl, carbonyl and epoxy [67]. The neutralization of bacteria by GO involves GO sheet adhesion, bacterial cell membrane piercing, phospholipid extraction and oxidative stress. Usually, larger GO sheets tend to wrap and isolate the bacterial cells, however, smaller GO sheets with sharp edges lead to profilation and subsequently phospholipid extraction. Hence, it can be concluded that smaller GO sheets with large number of defects exhibit good antibacterial activity [68].

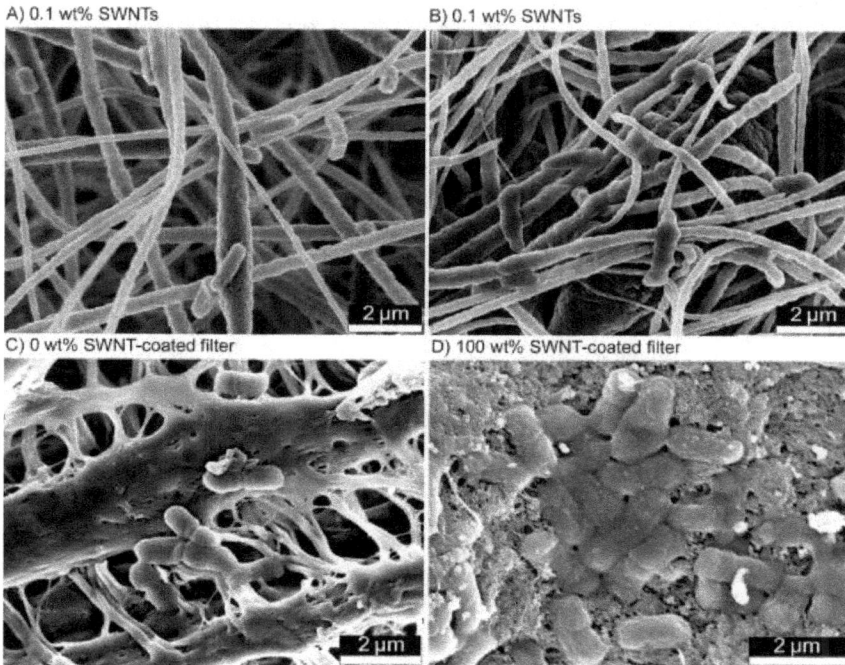

Figure 8.2 SEM micrographs displaying E. coli that have been incubated for 1 h on (A, B) polysulfone mats containing 0.1 wt % SWNTs and commercial filters coated with (C) 0 and (D) 100 wt % SWNT coating, respectively. Micrographs A and C display viable cells, whereas B and D display cells that have been inactivated by SWNTs. Reproduced from Reference 65 with permission from American Chemical Society.

Mural *et al.* [38] incorporated amine terminated GO in PE membranes for improving the mechanical strength and antibacterial performance against E. coli. In another study, the PVDF membranes functionalized with amine terminated graphene oxide quantum dots was reported by Zeng *et al.* [69], and the results revealed the biocidal nature of GOQDs-PVDF membranes, which inhibited the proliferation and biofilm formation effectively. The study revealed that functionalized PVDF membrane effectively eliminated both E.coli and S. aureus bacteria. Samantaray *et al.* [70] modified the surface of PVDF/PBSA (poly(butylene succinate-co-adipate)) membrane via

GO and GO anchored with trihexyltetradecylphosphonium chloride (TPCI), as shown in Figure 8.3. The TPCI anchored GO induced synergistic effect and resulted in enhanced antibacterial and antifouling properties. Mural *et al.* [71] prepared GO immobilized PE 3D porous polyolefin membranes for water treatment (Figure 8.4). The immobilized GO over PE membrane resulted in 99% elimination of the E. coli bacteria. The researchers also investigated the effect of reduced graphene oxide (rGO), Ag and rGO-Ag on the bactericidal properties of porous PE membranes. The study revealed that the presence of rGO-Ag produced synergistic effect and led to the improved bactericidal performance against E. coli (Figure 8.5) [72].

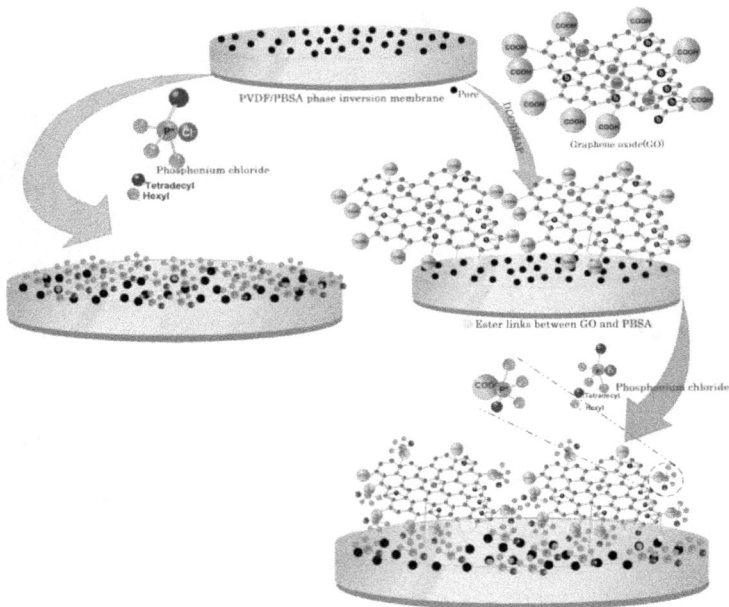

Figure 8.3 Surface modification of PVDF/PBSA phase inversion membrane. Reproduced from Reference 70 with permission from Elsevier.

8.4.3 Quaternary Ammonium Compounds (QACs)

QACs, with positively charged polyatomic ions structure, have been extensively used as antibacterial agents in polymeric membranes. The presence of positively charged structure in the quaternary ammonium compounds forms electrostatic interactions with negatively

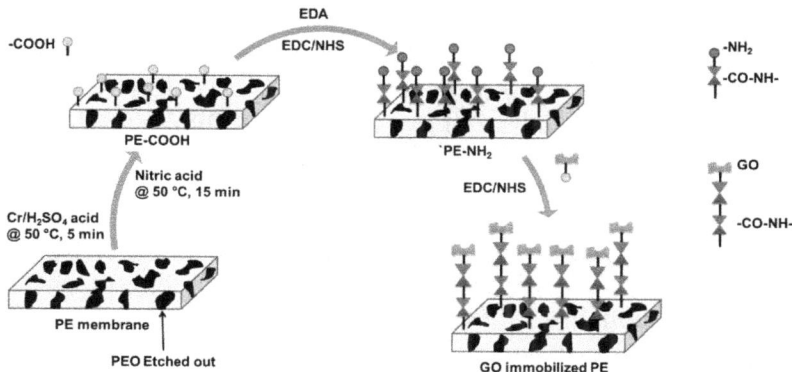

Figure 8.4 Surface treatment and immobilization of GO on token PE membranes. Reproduced from Reference 71 with permission from Royal Society of Chemistry.

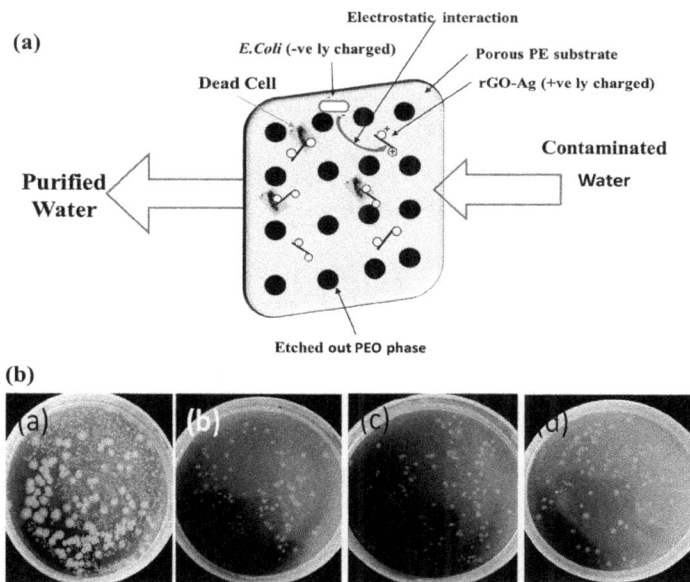

Figure 8.5 (a) A schematic of the the key role of rGO–Ag nanoparticles as bactericidal agent in porous PE membranes for water purification; (b) antibacterial properties of the composites shown on total agar plate counts. The number of bacterial colonies that appear on the agar plate, relative to the control, after 12 h of inoculation for 90/10 PE/PEO (a) neat blends (b) with 1 wt% rGO (c) with 1 wt% Ag and (d) with 1 wt% rGO–Ag. Reproduced from Reference 72 with permission from Royal Society of Chemistry.

charged bacterial cell wall and consequently results in the cell wall lysis, followed by disruption, protein denaturation and reduction in nutrients to the bacterial cells [73]. Yao *et al.* [74] fabricated microporous polypropylene (PP) hollow fiber membranes by grafting copolymer brushes of poly(ethylene glycol) monomethacrylate (PEGMA) and PDMAEMA poly (2- (dimethylamino) ethyl methacrylate)), followed by the quaternization of polymer brushes. The modified hydrophilic membranes showed good bactericidal activity.

Park *et al.* [75] prepared benzyl triethylammonium chloride (BTEAC) functionalized PVA nanofibers followed by deposition of the same on polycarbonate (PC) membranes. The membranes revealed extraordinary anti-biofouling and antibacterial properties. Ni *et al.* [76] fabricated membranes with excellent antibacterial activity by coating terpolymer consisting of poly(methylacryloxyethyldimethyl benzyl ammonium chloride-r-acrylamide-r-2-hydroxylethylmethacrylate) (P(MDBAC-r-Am-rHEMA)) over commercially available RO membrane. The modified membrane exhibited high flux in comparison to pristine membrane. Zhang *et al.* [77] synthesized composite PVDF microflitration membranes using QAC@carbon, which effectively decreased the microbial growth.

8.4.4 Silver based Nanocomposites

The studies have shown that silver nanoparticles (AgNPs) exhibit bactericidal effect even at very low concentration [53]. Various silver based compounds have been used for the fabrication of composite membranes like silver nitrate, silver tetrafluoroborate, silver nitrate, etc. The neutralization of E. coli by silver can be attributed to its leaching [78]. For instance, the silver nanoparticles incorporated in polysulfone membranes significantly inhibited the bacterial growth, and the extent of leaching was under the WHO permissible limit [79]. Sandwich membranes by blending AgNPs, halloysite nanotubes (HNTs) and rGO into PES matrix were prepared by Zhao *et al.* [80]. Efficient dispersion of AgNPs within the matrix resulted in higher hydrophilicity, improved flux and surface smoothness. The study revealed ideal bacteriostatic ability of the mixed matrix membranes (MMM) against E. coli.

Sharma *et al.* [81] studied the significance of the surface assembly of biocidal nanoparticles on PVDF membranes for controlled leaching of nanoparticles. In this study, PVDF membrane was modified with thiol groups and biocidal silver nanoparticles were decorated over the thiolated PVDF membranes, as shown in Figure 8.6. In another study, the effect of adding different types of nanoparticles like TiO_2, Ag, Ag-CNTs, TiO_2-CNTs, Ag-TiO_2 and Ag@TiO_2-CNTs to the melt mixed 80/20 PVDF/PMMA blends was analyzed [82]. The study demonstrated that the blends with Ag@TiO_2-CNTs showed improved antibacterial and antifouling properties in comparison to other blends. The antibacterial properties of the porous membranes are depicted schematically in Figure 8.7.

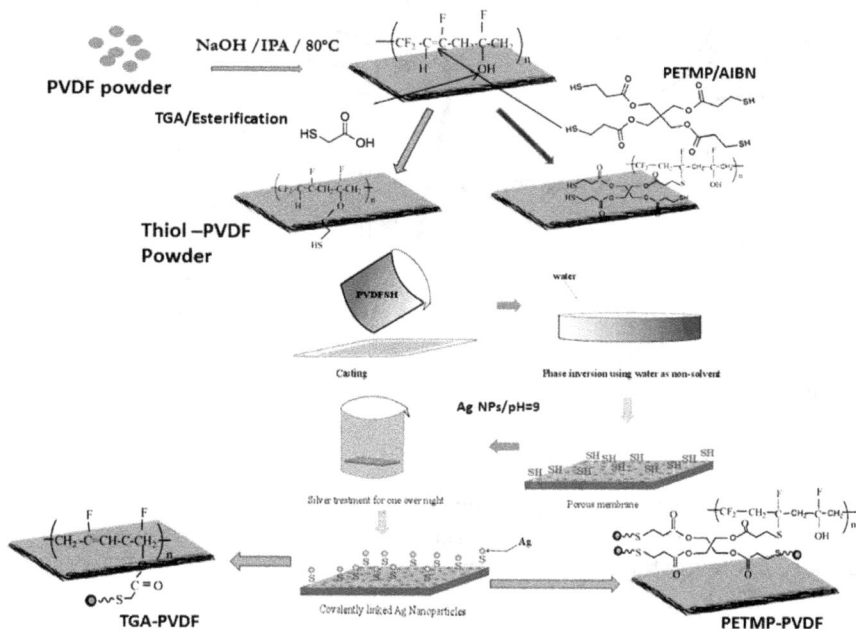

Figure 8.6 Schematic illustration of the fabrication of silver nanoparticles covalently bound on the PVDF membrane *via* esterification and thiol-ene reactions. Reproduced from Reference 81 with permission from Royal Society of Chemistry.

Mural *et al.* [83] prepared antibacterial porous PE membranes with controlled leaching of biocidal silver ions. The leaching of silver in the final permeate was found to be lower than the permissible limit, thus, opening the route to antibacterial polyolefin-based membranes for water decontamination Figure 8.8 shows the enhanced antibacterial performance of the PE membrane decorated with silver ions. The authors also synthesized Surlyn (partially neutralized zinc salt of polyethylene) based porous membranes for enhancing the antifouling and antibacterial properties of the porous membranes, developed by decorating silver ions on the on polyethylene ionomer membranes (Figure 8.9) [84].

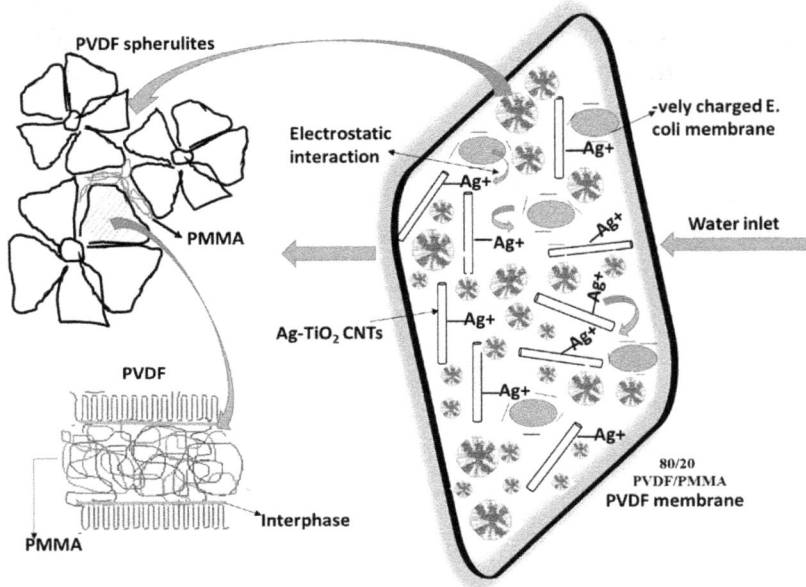

Figure 8.7 Schematic showing the antibacterial properties of porous membranes. Reproduced from Reference 82 with permission from Royal Society of Chemistry.

Yuksel *et al.* [85] studied the influence of silver nanoparticles on the antibacterial mechanism of membranes, prepared with different polymers, such as polystyrene (PS), PES and cellulose acetate (CA). Xu *et al.* [86] fabricated the L-dopa crosslinked mixed matrix membranes made of PSF and silver-copper oxide wires. The attachment of silver-copper oxide wires through covalent bonds reduced the leaching of silver ions. Huang *et al.* [87] studied the antibacterial efficiency of immobilized AgNPs on PSF membranes, developed by PDA mediated *in-situ* reduction of silver ammonia aqueous solution. In another study, PES membrane impregnated with silver nanoparticles was prepared by Zodrow *et al.* [88]. The authors found that PES membrane modified with 0.9 wt% nanoparticles greatly reduced the growth of E. coli bacteria over membrane. However, the leaching of the silver ions from the membrane surface was one of the limitations of this approach, which consequently decreased the performance of the membranes with time.

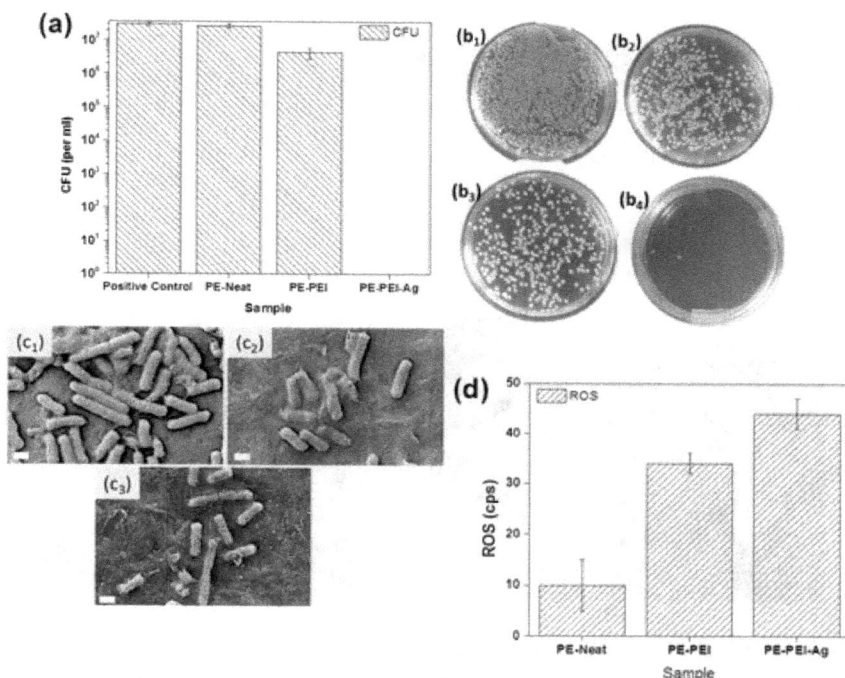

Figure 8.8 (a) Colony forming unit (CFU) per mL of blank (without PE membranes), PE-Neat, PE-PEI and PE-PEI-Ag membranes, (b) The agar plates of bacterial colonies after 12 h of inoculation of blank (without PE membranes) (b$_1$), PE-Neat (b$_2$), PE-PEI (b$_3$) and PE-PEI-Ag (b$_4$), (c) SEM morphology of Neat PE (c$_1$), PE-PEI (c$_2$) and PE-PEI–Ag (c$_3$) on which bacterial colonies were grown and (d) ROS intensity (at 528 nm) of PE membranes with culture and dye. Reproduced from Reference 83 with permission from Wiley.

In summary, the inclusion of inorganic nanoparticles into polymeric membranes significantly improves their antibacterial and antifouling properties, however, this strategy still suffers from some limitations. For instance, agglomeration of nanoparticles at high concentration is a major drawback associated with this approach. Hence, the studies should be focused on adding the optimum concentration of the nanoparticles so that the required properties of the polymeric membranes can be achieved.

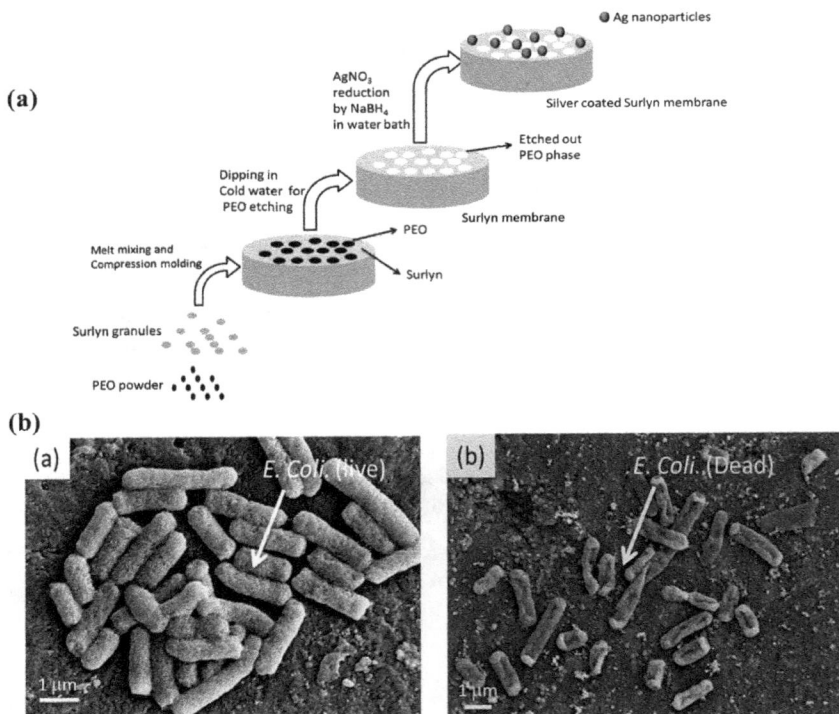

Figure 8.9 (a) Illustration of the membrane preparation, selective etching and silver coating on the token membranes; (b) SEM morphology of Surlyn membrane (a) and with 10 wt/vol Ag coating (b). Reproduced from Reference 84 with permission from Royal Society of Chemistry.

8.4.5 Zwitterionic Polymers

The zwitterionic polymers consist of equal number of cationic and anionic groups within the same monomeric unit. The zwitterionic polymers have been used in the recent years to functionalize the membrane surface for enhancing both antifouling and antibacterial properties. The functionalization of the membrane surface using zwitterionic polymers imparts hydrophilicity and, thus, significantly reduces the fouling of the membrane via anti-adhesion behavior. Several approaches such as surface initiated atomic transfer radical polymerization, UV graft polymerization, redox initiated graft polymerization, etc., have been used to functionalize the membranes with zwitterionic polymers [89]. In order to achieve excellent antibacterial properties, Zhao *et al.* [90] fabricated PES membranes by surface quaternization from a reactive PES based copolymer additive [poly(ether sulfone)-block-poly(2-(dimethylamino) ethyl methacrylate) (PDMAEMA-b-PES-b-PDMAEMA)] with excellent antibacterial nature against both gram positive and gram negative bacteria (Figure 8.10). The work opens new avenues for further functionalization of PDMAEMA blended membranes. Though, the presence of zwitterions polymers on the membrane surface resulted in reducing the biofouling of the membranes, however, long term stability using this approach was not been investigated in detail.

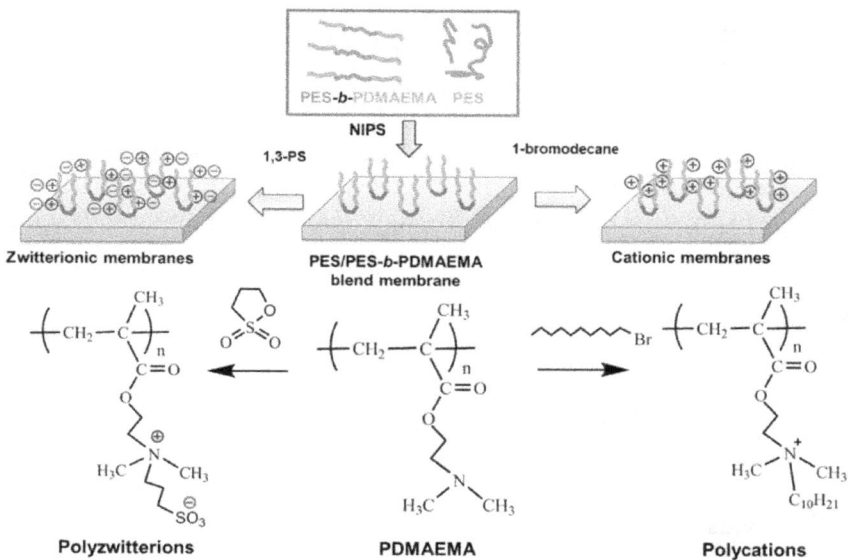

Figure 8.10 Schematic of the fabrication and surface quaternization of the PES/PDMAEMA-*b*-PES-*b*-PDMAEMA blend membranes. Reproduced from Reference 90 with permission from American Chemical Society.

8.5 Summary and Outlook

In conclusion, there is a significant demand for the fabrication of sustainable polymeric membranes with excellent antifouling and antibacterial properties for use in seawater treatment and contaminated water purification. Membrane fouling mechanisms are complex, owing to the different interactions between foulants and membrane surface. In order to enhance the antifouling and antibacterial characteristics of the membranes, several strategies have been employed by incorporating various antifouling and antibacterial agents. The inclusion of hydrophilic polymers, zwitterions, inorganic nanoparticles and antimicrobial polymers has been observed to significantly improve both antifouling and antibacterial properties. The incorporation of inorganic nanoparticles like silver, TiO_2 and GO improve both mechanical and antibacterial properties of polymeric membranes. However, leaching of nanoparticles is one of the major problems faced in this strategy, which can lead to the contamination of the membrane environment. Though a variety of antifouling and antibacterial membranes synthesized at laboratory scale have shown significant potential, however, only a few approaches can be transferred from the laboratory scale to the industrial level, due to harsh modification conditions and high cost. Hence, developing an easily scalable and more controllable method to fabricate membranes with multifunctional properties is of great significance. Besides, cooperative contributions are needed from different fields like chemistry, materials science, biology and environmental science while fabricating the membranes.

Acknowledgements

SP would like to acknowledge DST for Women Scientist-B under scheme no. SR/WOS-B/565/2016.

References

1. Elimelech, M. (2006) The global challenge for adequate and safe water. *Journal of Water Supply: Research and Technology - AQUA*, **55**(1), 3-10.
2. Shannon, M. A., Bohn, P. W., Elimelech, M., Georgiadis, J. G., Marinas, B. J., and Mayes, A. M. (2008) Science and technology for water purification in the coming decades. *Nature*, **452**(7185), 301-310.
3. Eliasson, J. (2015) The rising pressure of global water shortages. *Nature*, **517**(7532), 6.
4. Tao, T., and Xin, K. (2014) A sustainable plan for China's drinking water: tackling pollution and using different grades of water for different tasks is more efficient than making all water potable. *Nature*, **511**(7511), 527-529.
5. Werber, J. R., Deshmukh, A., and Elimelech, M. (2016) The critical need for increased selectivity, not increased water permeability,

for desalination membranes. *Environmental Science and Technology Letters*, **3**(4), 112-120.

6. Zhuang, Y., Ren, H., Geng, J., Zhang, Y., Zhang, Y., Ding, L., and Xu, K. (2015) Inactivation of antibiotic resistance genes in municipal wastewater by chlorination, ultraviolet, and ozonation disinfection. *Environmental Science and Pollution Research*, **22**(9), 7037-7044.

7. Mukherjee, M., and De, S. (2018) Antibacterial polymeric membranes: a short review. *Environmental Science: Water Research and Technology*, **4**, 1078-1104.

8. Mi, B., and Elimelech, M. (2008) Chemical and physical aspects of organic fouling of forward osmosis membranes. *Journal of Membrane Science*, **320**(1-2), 292-302.

9. Zhang, R., Liu, Y., He, M., Su, Y., Zhao, X., Elimelech, M., and Jiang, Z. (2016) Antifouling membranes for sustainable water purification: strategies and mechanisms. *Chemical Society Reviews*, **45**(21), 5888-5924.

10. Zodrow, K. R., Tousley, M. E., and Elimelech, M. (2014) Mitigating biofouling on thin-film composite polyamide membranes using a controlled-release platform. *Journal of Membrane Science*, **453**, 84-91.

11. Lin, R., Hernandez, B. V., Ge, L., and Zhu, Z. (2018) Metal organic framework based mixed matrix membranes: an overview on filler/polymer interfaces. *Journal of Materials Chemistry A*, **6**(2), 293-312.

12. Le-Clech, P., Chen, V., Fane, and T. A. (2006) Fouling in membrane bioreactors used in wastewater treatment. *Journal of Membrane Science*, **284**(1-2), 17-53.

13. Lee, S., Cho, J., and Elimelech, M. (2004) Influence of colloidal fouling and feed water recovery on salt rejection of RO and NF membranes. *Desalination*, **160**(1), 1-12.

14. Chen, W., Su, Y., Peng, J., Dong, Y., Zhao, X., Jiang, Z. (2011) Engineering a robust, versatile amphiphilic membrane surface through forced surface segregation for ultralow flux-decline. *Advanced Functional Materials*, **21**(1), 191-198.

15. Rana, D., and Matsuura, T. (2010) Surface modifications for antifouling membranes. *Chemical Reviews*, **110**(4), 2448-2471.

16. Wang, C., Li, Q., Tang, H., Yan, D., Zhou, W., Xing, J., and Wan, Y. (2012) Membrane fouling mechanism in ultrafiltration of succinic acid fermentation broth. *Bioresource Technology*, **116**, 366-371.

17. Juang, R.-S., Chen, H.-L., and Chen, Y.-S. (2008) Resistance - in-series analysis in cross-flow ultrafiltration of fermentation broths of Bacillus subtilis culture. *Journal of Membrane Science*, **323**(1), 193-200.

18. Wei, Q., Becherer, T., Angioletti-Uberti, S., Dzubiella, J., Wischke, C.,

Neffe, A. T., Lendlein, A., Ballauff, M., and Haag, R. (2014) Protein interactions with polymer coatings and biomaterials. *Angewandte Chemie, International Edition*, **53**(31), 8004-8031.

19. Huisman, I. H., Prádanos, P., and Hernández, A. (2000) The effect of protein–protein and protein–membrane interactions on membrane fouling in ultrafiltration. *Journal of Membrane Science*, **179**(1-2), 79-90.

20. Norde, W. (1994) Protein adsorption at solid surfaces: A thermodynamic approach. *Pure and Applied Chemistry*, **66**(3), 491-496.

21. Matin, A., Khan, Z., Zaidi, S., and Boyce, M. (2011) Biofouling in reverse osmosis membranes for seawater desalination: phenomena and prevention. *Desalination*, **281**, 1-16.

22. Magin, C. M., Cooper, S. P., and Brennan, A. B. (2010) Non-toxic antifouling strategies. *Materials Today*, **13**(4), 36-44.

23. Li, R., and Barbari, T. (1995) Performance of poly (vinyl alcohol) thin-gel composite ultrafiltration membranes. *Journal of Membrane Science*, **105**(1-2), 71-78.

24. Jiang, J., Zhu, L., Zhu, L., Zhang, H., Zhu, B., and Xu, Y. (2013) Antifouling and antimicrobial polymer membranes based on bioinspired polydopamine and strong hydrogen-bonded poly (N-vinyl pyrrolidone). *ACS Applied Materials and Interfaces*, **5**(24), 12895-12904.

25. Sagle, A. C., Ju, H., Freeman, B. D., and Sharma, M. M. (2009) PEG-based hydrogel membrane coatings. *Polymer*, **50**(3), 756-766.

26. Mauter, M. S., Wang, Y., Okemgbo, K. C., Osuji, C. O., Giannelis, E. P., and Elimelech, M. (2011) Antifouling ultrafiltration membranes via post-fabrication grafting of biocidal nanomaterials. *ACS Applied Materials and Interfaces*, **3**(8), 2861-2868.

27. Ben-Sasson, M., Zodrow, K. R., Genggeng, Q., Kang, Y., Giannelis, E. P., and Elimelech, M. (2013) Surface functionalization of thin-film composite membranes with copper nanoparticles for antimicrobial surface properties. *Environmental Science and Technology*, **48**(1), 384-393.

28. Baxamusa, S. H., Im, S. G., and Gleason, K. K. (2009) Initiated and oxidative chemical vapor deposition: a scalable method for conformal and functional polymer films on real substrates. *Physical Chemistry Chemical Physics*, **11**(26), 5227-5240.

29. Yang, R., Asatekin, A., and Gleason, K. K. (2012) Design of conformal, substrate-independent surface modification for controlled protein adsorption by chemical vapor deposition (CVD). *Soft Matter*, **8**(1), 31-43.

30. Ozaydin-Ince, G., Matin, A., Khan, Z., Zaidi, S. J., and Gleason, K. K. (2013) Surface modification of reverse osmosis desalination membranes by thin-film coatings deposited by initiated chemical vapor deposition. *Thin Solid Films*, **539**, 181-187.

31. Matin, A., Shafi, H., Khan, Z., Khaled, M., Yang, R., Gleason, K., and Rehman, F. (2014) Surface modification of seawater desalination reverse osmosis membranes: characterization studies and performance evaluation. *Desalination*, **343**, 128-139.

32. Shafi, H. Z., Khan, Z., Yang, R., and Gleason, K. K. (2015) Surface modification of reverse osmosis membranes with zwitterionic coating for improved resistance to fouling. *Desalination*, **362**, 93-103.

33. Rodriguez-Hernandez, J. (2017) *Polymers Against Microorganisms*, Springer, Germany.

34. Wang, Y.-Q., Su, Y.-L., Sun, Q., Ma, X.-L., and Jiang, Z.-Y. (2006) Generation of anti-biofouling ultrafiltration membrane surface by blending novel branched amphiphilic polymers with polyethersulfone. *Journal of Membrane Science*, **286**(1-2), 228-236.

35. Mansouri, J., Harrisson, S., and Chen, V. (2010) Strategies for controlling biofouling in membrane filtration systems: challenges and opportunities. *Journal of Materials Chemistry*, **20**(22), 4567-4586.

36. Idris, A., Zain, N. M., and Noordin, M. (2007) Synthesis, characterization and performance of asymmetric polyethersulfone (PES) ultrafiltration membranes with polyethylene glycol of different molecular weights as additives. *Desalination*, **207**(1-3), 324-339.

37. Liu, Y., Koops, G., and Strathmann, H. (2003) Characterization of morphology controlled polyethersulfone hollow fiber membranes by the addition of polyethylene glycol to the dope and bore liquid solution. *Journal of Membrane Science*, **223**(1-2), 187-199.

38. Mural, P. K. S., Banerjee, A., Rana, M. S., Shukla, A., Padmanabhan, B., Bhadra, S., Madras, G., and Bose, S. (2014) Polyolefin based antibacterial membranes derived from PE/PEO blends compatibilized with amine terminated graphene oxide and maleated PE. *Journal of Materials Chemistry A*, **2**(41), 17635-17648.

39. Jeon, S., Lee, J., Andrade, J., and De Gennes, P. (1991) Protein-surface interactions in the presence of polyethylene oxide: I. Simplified theory. *Journal of Colloid and Interface Science*, **142**(1), 149-158.

40. Kang, S., Asatekin, A., Mayes, A. M., and Elimelech, M. (2007) Protein antifouling mechanisms of PAN UF membranes incorporating PAN-g-PEO additive. *Journal of Membrane Science*, **296**(1-2), 42-50.

41. Gudipati, C. S., Finlay, J. A., Callow, J. A., Callow, M. E., and Wooley, K. L. (2005) The antifouling and fouling-release perfomance of hyperbranched fluoropolymer (HBFP)– poly (ethylene glycol)(PEG) composite coatings evaluated by adsorption of biomacromolecules and the green fouling alga Ulva. *Langmuir*, **21**(7), 3044-3053.

42. Pasmore, M., Todd, P., Smith, S., Baker, D., Silverstein, J., Coons, D., and Bowman, C. N. (2001) Effects of ultrafiltration membrane sur-

face properties on Pseudomonas aeruginosa biofilm initiation for the purpose of reducing biofouling. *Journal of Membrane Science*, **194**(1), 15-32.

43. Zhang, Y., Wang, Z., Lin, W., Sun, H., Wu, L., and Chen, S. (2013) A facile method for polyamide membrane modification by poly (sulfobetaine methacrylate) to improve fouling resistance. *Journal of Membrane Science*, **446**, 164-170.

44. Meng, J., Cao, Z., Ni, L., Zhang, Y., Wang, X., Zhang, X., and Liu, E. (2014) A novel salt-responsive TFC RO membrane having superior antifouling and easy-cleaning properties. *Journal of Membrane Science*, **461**, 123-129.

45. Li, G., Cheng, G., Xue, H., Chen, S., Zhang, F., and Jiang, S. (2008) Ultra low fouling zwitterionic polymers with a biomimetic adhesive group. *Biomaterials*, **29**(35), 4592-4597.

46. Khulbe, K., Feng, C., and Matsuura, T. (2010) The art of surface modification of synthetic polymeric membranes. *Journal of Applied Polymer Science*, **115**(2), 855-895.

47. Van der Bruggen, B. (2009) Chemical modification of polyethersulfone nanofiltration membranes: A review. *Journal of Applied Polymer Science*, **114**(1), 630-642.

48. Chen, Y., Ying, L., Yu, W., Kang, E., and Neoh, K. (2003) Poly (vinylidene fluoride) with grafted poly (ethylene glycol) side chains via the RAFT-mediated process and pore size control of the copolymer membranes. *Macromolecules*, **36**(25), 9451-9457.

49. Ye, G., Lee, J., Perreault, F. O., and Elimelech, M. (2015) Controlled architecture of dual-functional block copolymer brushes on thin-film composite membranes for integrated "defending" and "attacking" strategies against biofouling. *ACS Applied Materials And Interfaces*, **7**(41), 23069-23079.

50. Cao, X., Ma, J., Shi, X., and Ren, Z. (2006) Effect of TiO2 nanoparticle size on the performance of PVDF membrane. *Applied Surface Science*, **253**(4), 2003-2010.

51. Madaeni, S., and Ghaemi, N. (2007) Characterization of self-cleaning RO membranes coated with TiO2 particles under UV irradiation. *Journal of Membrane Science*, **303**(1-2), 221-233.

52. Yu, S., Zuo, X., Bao, R., Xu, X., Wang, J., and Xu, J. (2009) Effect of SiO2 nanoparticle addition on the characteristics of a new organic–inorganic hybrid membrane. *Polymer*, **50**(2), 553-559.

53. Zhu, J., Hou, J., Zhang, Y., Tian, M., He, T., Liu, J., and Chen, V. (2017) Polymeric antimicrobial membranes enabled by nanomaterials for water treatment. *Journal of Membrane Science*. **550**, 173-197.

54. Chatterjee, A. K., Chakraborty, R., and Basu, T. (2014) Mechanism of antibacterial activity of copper nanoparticles. *Nanotechnology*, **25**(13), 135101.

55. Xu, J., Feng, X., Chen, P., and Gao, C. (2012) Development of an anti

bacterial copper (II)-chelated polyacrylonitrile ultrafiltration membrane. *Journal of Membrane Science*, **413**, 62-69.

56. Chen, Y., Zhang, Y., Liu, J., Zhang, H., and Wang, K. (2012) Preparation and antibacterial property of polyethersulfone ultrafiltration hybrid membrane containing halloysite nanotubes loaded with copper ions. *Chemical Engineering Journal*, **210**, 298-308.

57. Raghupathi, K. R., Koodali, R. T., and Manna, A. C. (2011) Size-dependent bacterial growth inhibition and mechanism of antibacterial activity of zinc oxide nanoparticles. *Langmuir*, **27**(7), 4020-4028.

58. Li, M., Zhu, L., and Lin, D. (2011) Toxicity of ZnO nanoparticles to Escherichia coli: mechanism and the influence of medium components. *Environmental Science and Technology*, **45**(5), 1977-1983.

59. Al-Hinai, M. H., Sathe, P., Al-Abri, M. Z., Dobretsov, S., and Al-Hinai, A. T., Dutta, J. (2017) Antimicrobial activity enhancement of poly (ether sulfone) membranes by in situ growth of ZnO nanorods. *ACS Omega*, **2**(7), 3157-3167.

60. Chung, Y. T., Mahmoudi, E., Mohammad, A. W., Benamor, A., Johnson, D., and Hilal, N. (2017) Development of polysulfone-nanohybrid membranes using ZnO-GO composite for enhanced antifouling and antibacterial control. *Desalination*, **402**, 123-132.

61. Damodar, R. A., You, S.-J., and Chou, H.-H. (2009) Study the self-cleaning, antibacterial and photocatalytic properties of TiO_2 entrapped PVDF membranes. *Journal of Hazardous Materials*, **172**(2-3), 1321-1328.

62. Rahimpour, A., Jahanshahi, M., Rajaeian, B., and Rahimnejad, M. (2011) TiO2 entrapped nano-composite PVDF/SPES membranes: Preparation, characterization, antifouling and antibacterial properties. *Desalination*, **278**(1-3), 343-353.

63. Pant, H. R., Pandeya, D. R., Nam, K. T., Baek, W.-i., Hong, S. T., and Kim, H. Y. (2011) Photocatalytic and antibacterial properties of a TiO_2/nylon-6 electrospun nanocomposite mat containing silver nanoparticles. *Journal of Hazardous Materials*, **189**(1-2), 465-471.

64. Kang, S., Pinault, M., Pfefferle, L. D., and Elimelech, M. (2007) Single-walled carbon nanotubes exhibit strong antimicrobial activity. *Langmuir*, **23**(17), 8670-8673.

65. Schiffman, J. D., and Elimelech, M. (2011) Antibacterial activity of electrospun polymer mats with incorporated narrow diameter single-walled carbon nanotubes. *ACS Applied Materials and Interfaces*, **3**(2), 462-468.

66. Rusen, E., Mocanu, A., Nistor, L. C., Dinescu, A., Calinescu, I., Mustățea, G., Voicu, S. t. I., Andronescu, C., and Diacon, A. (2014) Design of antimicrobial membrane based on polymer colloids/multiwall carbon nanotubes hybrid material with silver nanoparticles. *ACS Applied Materials and Interfaces*, **6**(20), 17384-17393.

67. Tian, T., Shi, X., Cheng, L., Luo, Y., Dong, Z., Gong, H., Xu, L., Zhong, Z., Peng, R., and Liu, Z. (2014) Graphene-based nanocomposite as an effective, multifunctional, and recyclable antibacterial agent. *ACS Applied Materials and Interfaces*, **6**(11), 8542-8548.

68. Perreault, F., De Faria, A. F., and Elimelech, M. (2015) Environmental applications of graphene-based nanomaterials. *Chemical Society Reviews*, **44**(16), 5861-5896.

69. Zeng, Z., Yu, D., He, Z., Liu, J., Xiao, F.-X., Zhang, Y., Wang, R., Bhattacharyya, D., and Tan, T. T. Y. (2016) Graphene oxide quantum dots covalently functionalized PVDF membrane with significantly-enhanced bactericidal and antibiofouling performances. *Scientific Reports*, **6**, 20142.

70. Samantaray, P. K., Madras, G., and Bose, S. (2018) PVDF/PBSA membranes with strongly coupled phosphonium derivatives and graphene oxide on the surface towards antibacterial and antifouling activities. *Journal of Membrane Science*, **548**, 203-214.

71. Mural, P. K. S., Jain, S., Kumar, S., Madras, G., and Bose, S. (2016) Unimpeded permeation of water through biocidal graphene oxide sheets anchored on to 3D porous polyolefinic membranes. *Nanoscale*, **8**(15), 8048-8057.

72. Mural, P. K. S., Sharma, M., Shukla, A., Bhadra, S., Padmanabhan, B., Madras, G., and Bose, S. (2015) Porous membranes designed from bi-phasic polymeric blends containing silver decorated reduced graphene oxide synthesized via a facile one-pot approach. *RSC Advances*, **5**(41), 32441-32451.

73. Peng, Q., Lu, S., Chen, D., Wu, X., Fan, P., Zhong, R., and Xu, Y. (2007) Poly (vinylidene fluoride)-graft-poly (N-vinyl-2-pyrrolidone) copolymers prepared via a RAFT-mediated process and their use in antifouling and antibacterial membranes. *Macromolecular Bioscience*, **7**(9-10), 1149-1159.

74. Yao, F., Fu, G.-D., Zhao, J., Kang, E.-T., and Neoh, K. G. (2008) Antibacterial effect of surface-functionalized polypropylene hollow fiber membrane from surface-initiated atom transfer radical polymerization. *Journal of Membrane Science*, **319**(1-2), 149-157.

75. Park, J.-A., and Kim, S.-B. (2017) Antimicrobial filtration with electrospun poly (vinyl alcohol) nanofibers containing benzyl triethylammonium chloride: Immersion, leaching, toxicity, and filtration tests. *Chemosphere*, **167**, 469-477.

76. Ni, L., Meng, J., Li, X., and Zhang, Y. (2014) Surface coating on the polyamide TFC RO membrane for chlorine resistance and antifouling performance improvement. *Journal of Membrane Science*, **451**, 205-215.

77. Zhang, X., Ma, J., Tang, C. Y., Wang, Z., Ng, H. Y., and Wu, Z. (2016) Antibiofouling polyvinylidene fluoride membrane modified by quaternary ammonium compound: direct contact-killing versus

induced indirect contact-killing. *Environmental Science and Technology*, **50** (10), 5086-5093.

78. Kim, J. H., Park, S. M., Won, J., and Kang, Y. S. (2005) Unusual separation property of propylene/propane mixtures through polymer/silver complex membranes containing mixed salts. *Journal of Membrane Science*, **248** (1-2), 171-176.

79. Taurozzi, J. S., Arul, H., Bosak, V. Z., Burban, A. F., Voice, T. C., Bruening, M. L., and Tarabara, V. V. (2008) Effect of filler incorporation route on the properties of polysulfone–silver nanocomposite membranes of different porosities. *Journal of Membrane Science*, **325** (1), 58-68.

80. Zhao, Q., Hou, J., Shen, J., Liu, J., and Zhang, Y. (2015) Long-lasting antibacterial behavior of a novel mixed matrix water purification membrane. *Journal of Materials Chemistry A*, **3**(36), 18696-18705.

81. Sharma, M., Padmavathy, N., Remanan, S., Madras, G., and Bose, S. (2016) Facile one-pot scalable strategy to engineer biocidal silver nanocluster assembly on thiolated PVDF membranes for water purification. *RSC Advances*, **6**(45), 38972-38983.

82. Sharma, M., Madras, G., and Bose, S. (2015) Unique nanoporous antibacterial membranes derived through crystallization induced phase separation in PVDF/PMMA blends. *Journal of Materials Chemistry A*, **3**(11), 5991-6003.

83. Mural, P. K. S., Jain, S., Madras, G., and Bose, S. (2017) Antibacterial Membranes for Water Remediation with Controlled Leaching of Biocidal Silver Aided by Prior Grafting of Poly (ethylene imine) on to Ozone-Treated Polyethylene. *ChemistrySelect*, **2**(2), 624-631.

84. Mural, P. K. S., Jain, S., Madras, G., Bose, S. (2016) Improving antifouling ability by site-specific silver decoration on polyethylene ionomer membranes for water remediation: assessed using 3D micro computed tomography, water flux and antibacterial studies. *RSC Advances*, **6**(91), 88057-88065.

85. Sile-Yuksel, M., Tas, B., Koseoglu-Imer, D. Y., and Koyuncu, I. (2014) Effect of silver nanoparticle (AgNP) location in nanocomposite membrane matrix fabricated with different polymer type on antibacterial mechanism. *Desalination*, **347**, 120-130.

86. Xu, Z., Ye, S., Zhang, G., Li, W., Gao, C., Shen, C., and Meng, Q. (2016) Antimicrobial polysulfone blended ultrafiltration membranes prepared with Ag/Cu$_2$O hybrid nanowires. *Journal of Membrane Science*, **509**, 83-93.

87. Huang, L., Zhao, S., Wang, Z., Wu, J., Wang, J., and Wang, S. (2016) In situ immobilization of silver nanoparticles for improving permeability, antifouling and anti-bacterial properties of ultrafiltration membrane. *Journal of Membrane Science*, **499**, 269-281.

88. Zodrow, K., Brunet, L., Mahendra, S., Li, D., Zhang, A., Li, Q., and Alvarez, P. J. (2009) Polysulfone ultrafiltration membranes impreg-

nated with silver nanoparticles show improved biofouling resistance and virus removal. *Water Research*, **43**(3), 715-723.

89. Ren, P.-F., Fang, Y., Wan, L.-S., Ye, X.-Y., and Xu, Z.-K. (2015) Surface
modification of polypropylene microfiltration membrane by grafting poly (sulfobetaine methacrylate) and poly (ethylene glycol):
Oxidative stability and antifouling capability. *Journal of Membrane
Science*, **492**, 249-256.

90. Zhao, Y.-F., Zhu, L.-P., Jiang, J.-H., Yi, Z., Zhu, B.-K., and Xu, Y.-Y.
(2014) Enhancing the antifouling and antimicrobial properties of
poly (ether sulfone) membranes by surface quaternization from a
reactive poly (ether sulfone) based copolymer additive. *Industrial
and Engineering Chemistry Research*, **53**(36), 13952-13962.

Index

V

W

Z